Mark,

I hope you enjoy the book. I hope you enjoy the book. Dad was a wonderful person and a very loyal American.

Regards,

Kep

Without Fear or Favor

Gulf Publishing Company
Book Division
Houston, London, Paris, Tokyo

Without Fear or Favor

The Story of
Charles Henry Keplinger

By
H. F. (Kep) Keplinger

Without Fear or Favor

**Library of Congress Cataloging in
Publication Data**

Keplinger, H. F. (Henry Ferdinand)
 Without fear or favor.
 Includes index.
 1. Keplinger, Charles Henry, 1910-1981.
 2. Petroleum engineers—United States—Biog-
raphy. I. Title.
TN140.K46K46 1982 622'.338'0924 [B] 82-11883

ISBN 0-87201-917-9

Book design by David Price

There's an old saying that aptly describes Charles Henry Keplinger's attitude about his profession: "Do the work without fear or favor, and let the chips fall where they may." Henry called 'em like he saw 'em all his life, and neither power nor friendship nor wealth could influence him when he had a professional decision to make . . . .

Michel T. Halbouty
Houston, Texas

Without Fear or Favor

CHAPTER ONE

Grass-rustling spring had come to central Kansas, bringing a balmy April Sabbath to the community of Lyons. The bright day had lured the faithful to worship in numbers to warm a pastor's heart, but now the last prayer had been offered, the final hymn sung. In prospect were platters of fried chicken, bowls of mashed potatoes and smoking vegetables, biscuits, and tall glasses of iced tea. But before these delights could be enjoyed at home, another ritual had to be observed—the arrival and departure of the mid-day Santa Fe "milk train." In 1913, in small towns all across America, "going to see the train come in" was an exciting Saturday and Sunday pastime for those unable to witness the spectacle on weekdays because of work, school, or household chores.

This train was no whistling monster hurtling past the depot for faraway places, the faces of its passengers and crew mere blurs of color. It was a local, a leisurely old coal-burner that seemed to wander around aimlessly, picking up freight and short-ride fares. But it was an Iron Horse nonetheless, and as the powerful engine throbbed at leash, emitting hissing clouds of steam, it was the next best thing to the Ringling Brothers, Barnum and Bailey Circus train that inched through town annually on its journey from and to more impressive stops along the route.

In the crowd at the station was a small boy, standing alone, wearing his Sunday starched shirt and rompers. He was a brown-haired boy with bright blue eyes set in a taut, expectant face. His name was Charles Henry Keplinger. Most grownups called him "Little Kep." He was not quite three years old.

Inside the station his father, Henry Rosborough Keplinger, was working hurriedly to finish his Sunday stint as railway clerk. He had brought his son to the station this day because his wife, Edna, was expecting their second child. He was anxious to get home. Even now the doctor was at his wife's bedside. Working swiftly, he paid no heed when the groaning engine puffed away from the station with its short string of boxcars.

Suddenly a door was thrown open and a man shouted, "Hank! Your boy's on that train!"

Keplinger threw down his pencil and rushed outside on the platform. The train was moving quickly across the prairie, but Keplinger could clearly see his son. The boy was on the cowcatcher, at the front of the engine. He clung to the engine with one hand while with the other he waved a happy goodbye to his father, the crowd on the platform, and the community of Lyons. "Little Kep" was off to see the world!

Willing hands helped the father place a handcar on the tracks. Elbows churning, spurred by worry and anger, Keplinger set out in desperate pursuit of the train. He guessed that someone in the crowd would telephone or telegraph ahead to halt the train at the next station, but Keplinger was driven by fear that the boy would lose his grip on the engine and fall beneath the wheels. So he propelled the handcar with all the strength his mind and muscles could muster, hoping that someone in the caboose or engine would see him before he fell too far behind.

And just when his strength began to falter, when his palms were blistering on the pump handles, the train began to slow down. Someone *had* seen him. The train stopped. It began backing toward him, and Keplinger coasted to meet it. He leaped from the handcar before it had stopped completely and ran down the length of the train, ignoring those who called out to him. "Charles Henry!" he was shouting. "Charles Henry!"

Charles Henry, not quite three, stepped down from the cowcatcher and onto the road bed. The incipient world traveler was grave of countenance, and his lower lip quivered. He had not been frightened during his bold adventure; indeed, he had loved it! But the sight of his advancing father was another matter. The boy read only displeasure in the urgency of Henry

Keplinger's approach and in the frown that creased his forehead. He could not know the great relief that surged through the man as he swept up the boy in his arms.

"All right, you men," the engineer sang out. "Get Henry and the boy aboard and get that handcar out of the way. We're going to take them back to Lyons."

Back to Lyons they went. And at home that afternoon, as Charles Henry was seated in a corner—as much to get him out of the way as to punish him—his brother Gilbert Suffield Keplinger was born. The train ride and Gilbert's birth would become a part of the family folklore

But the ride was significant in itself. It was symbolic. It was symbolic because for much of his life, as the premier energy consultant, Charles Henry Keplinger would be waving goodbye with a smile at stops around the globe: from a remote oil camp in northern India to the most fashionable hotel in Paris. From Quito to Quebec, Texarkana to Tierra del Fuego. St. Moritz, Rome, and Belgrade. Algiers. Tehran. London, Ankara, Sofia, and Long Beach, California. El Centro, Colombia, and Maracaibo, Venezuela. Saskatchewan and Siberia. Houston and Hombre Pintada. And everywhere else on the globe men would drill for, produce, refine, ship or even just talk about oil and gas, coal and lignite, helium and geysers, primary, secondary, and tertiary recovery of hydrocarbons. And he would wave goodbye from his home and office in Tulsa, Oklahoma, his happy smile promising that he would always be delighted to get back home again.

☐

Both Henry Rosborough Keplinger and Edna Marilla Suffield were of pioneer Kansas stock. Indeed, Keplinger could trace his lineage to medieval Scotland and Ireland through the Rosborough branch of his family. His first American antecedents were Alexander Rosborough and Martha Gaston, Alexander's French Huguenot wife, who came to South Carolina from Ireland in 1769. Some of their descendants migrated to Indiana and Illinois. Eventually, their great granddaughter, Martha Elvira Rosborough, married Thomas Keplinger, son of Noah Keplinger, who had moved to Illinois from Pennsylvania.

Seven years after their marriage, when the United States government opened up Kansas for homesteading in 1879, Thomas and Martha Keplinger joined in the great migration.

After a three-week journey in a covered wagon, the couple settled in Greenwood County. They reared five children to maturity; Henry Rosborough was their only son.

In 1908 Henry married Edna Marilla Suffield, daughter of Charles and Nora Suffield, who had been among the first to settle McPherson County, Kansas. Charles Henry, the cowcatcher rider, was born June 7, 1910, in Lyons.

Henry Rosborough Keplinger was called a "good-looking man," and his pictures verify that description. He had an easygoing manner that endeared him to both sexes of all ages. Edna, an attractive woman, had been a schoolteacher before their marriage. She had been a disciplinarian with a high regard for excellence. When she forsook the classroom, she brought these characteristics into her home, and with them her strong Presbyterian faith.

She would live with Henry Rosborough Keplinger and bear his children in apparent contentment, but it is likely that she was disappointed that he did not share her vaulting ambition for him. Keplinger worked hard at his job. He supported his family in good times and bad. But he must have felt that his horizons were limited, and he did not dream beyond them.

So Edna turned to her first born. She had realized early that Charles Henry was a precocious child, and now she set about molding him to her mind's desire. Not that she failed to shower her affection on Gilbert, and then on Thomas Earl, who was to come along five years later. But she firmly guided Charles Henry toward the unquestioned blessings of hard work and scholarship. And she taught him thrift.

Fortunately for them both, and particularly for Charles Henry, nature had granted the boy a propensity to succeed. He was an eager pupil, one to whom the dreariest drudgery was but the means to an end, and a nickel a treasure in itself.

And in 1915, when he was five years old, something occurred that was to set the pattern for his life. About two miles northwest of El Dorado, in southeast Kansas, the Wichita Natural Gas Company, later Empire Gas and Fuel Company and still later Cities Service Oil Company, brought in Stapleton #1 to set off one of America's greatest oil booms.

El Dorado, a town of fewer than 4,000 inhabitants, almost immediately swelled to a population of 20,000. And Cities Service built a town of its own as the field blossomed: a bustling community of 2,500 or so, complete with company-owned housing, swimming pools, recreation halls, drugstores, grocery stores—even a police force and a fire department. It was called Oil Hill.

The Sante Fe transferred Henry Rosborough Keplinger to El Dorado to help handle the tremendous rail traffic the boom engendered. Thus Charles Henry Keplinger was to spend his formative years with the smells of oil and gas in his nostrils. His early heroes were the swaggering drillers whose iron-shod boots cracked like pistol shots on the El Dorado sidewalks.

He started to school in El Dorado at the age of seven, and he also started to work. His father built him a shoe-shine box of heavy wood. After school the boy would race home, get his box, then hurry downtown. There on a busy street corner he would begin his sing-song chant, "Shine 'em up, a nickel! Shine 'em up, a nickel!" Kneeling on the sidewalk, he spread his polish on boots and shoes, buffed them with his brush, and made them gleam with his flashing rag. Many a boomer, warmed by the boy's smile, dropped a dime—and occasionally a quarter—into the outstretched palm.

When the weather was bad, he swept out speakeasies and cafes. He filled in as a potato peeler at a small Central Avenue restaurant when the regular peeler succumbed to cheap gin. As he grew older, he also delivered milk in the early morning hours before he went to school.

By the time he was nine he had learned about golf—as the smallest Saturday and Sunday caddy at the El Dorado Country Club course and later at the American Legion greens. "He was so young and small," an oldtimer recalled, "that players hesitated to take him out. So he was always the last caddy to get hired. And on the last holes he was always dragging the bag, not carrying it. But in a year or two he got bigger and stronger, and Coxey Evans, who was commander of the American Legion Post, got to where he'd pick him first off the bench."

Coxey Evans also was a toolpusher at Oil Hill, and several years later, shortly before Charles Henry entered his teens, Evans got him a job as a roustabout in the field during the summer months. Charles Henry loaded and unloaded trucks and carried buckets of ice water to thirsty roughnecks who called out, "Waterjack! You oughta been here and half-way back!" How he longed then to be a driller, standing on the derrick floor and sending the spinning bit thousands of feet beneath the earth's surface to the pay sand. He fell in love with the machinery, and he studied the pumps and engines and boilers like a scientist who had stumbled upon the artifacts of a forgotten generation.

With all of this work, he had not neglected his schooling. He soaked up knowledge like a biscuit in red-eye gravy. He had no time for after-school sports; his time for play was at school recess periods. Before he entered El Dorado High School, he told Edna he wanted to be an engineer and work in the oil industry, that he would continue working and saving his money so that he could attend the University of Kansas.

"Physics, Charles Henry," Edna told him. "Take every class in physics they offer in high school." She had listened intently when he had told her about what he had observed at Oil Hill. "Everything in the oil business relates to physics," she said. "Mechanics, the design of the equipment they use, the flow of fluids, other things I don't understand. So you concentrate on physics."

By the spring of 1923, Henry, as he was now called by everyone except his mother, had saved $788.32, an amazing amount for a boy to amass in six years. Every week he had dutifully placed the money earned in the Butler County State Bank. At the rate he was earning and saving, he and Edna calculated he could well afford to attend the University of Kansas when he completed high school. At this time, the El Dorado area was still booming; almost every day there were Page One newspaper stories about new wells being brought in.

But on March 30, 1923, the Butler County State Bank shut its doors. Examiners said its assets were not sufficient to balance off its outstanding debts.

Henry could not at first allow himself to believe that the fruits of six years of hard work had vanished. And there was something else: the bank had been an institution, an integral part of the adult world. He was suffering his first genuine loss of faith because the institution had reneged on the compact between them.

Gilbert was suffering, too. He had been working in Henry's footsteps, and he had lost $150 in the bank closing. The parents tried to soothe their sons, but the hurt would stay with Henry all of his life. In his last years he would describe the bank closing as the most traumatic incident of his youth. He would remember his loss to the exact penny—$788.32.

☐

He found the remedy for his psychic pain: he simply worked harder and longer and studied with greater intensity. Edna knew that school was not all study and classrooms. She encouraged Henry to participate in school activities, and he took her advice—as long as the activities didn't interfere with his work.

It was Edna's mother, Mrs. Nora Suffield of Canton, Kansas, the McPherson County pioneer, who set young Henry on the path to becoming the social creature she sensed he longed to be. She bought him a trumpet. "Get in the band with those other high school youngsters," she admonished her grandson. "Mix with people!"

So Henry took the trumpet to school and learned to play it. He joined the school band and the orchestra, and someone told him he could make money by playing with the Butler County Band. At 14 he became the band's youngest member, and he would blow his horn all around Kansas in subsequent years, playing at fairs and for other festivities.

Edna bought Gilbert a violin and arranged for him to take lessons from a competent instructor. Gilbert attended the violin classes, but with great reluctance. One class day he searched his room for his violin but couldn't find it. Later in the evening he learned that Henry had sold it—with Edna's permission—for more money than Edna had paid for it. It was Henry's belief that Gilbert wasn't interested in the violin, and that he couldn't play it very well.

"Perhaps he was right," Gilbert said with a laugh many years later, "but I could play the violin as well as he could play that trumpet!"

Obeying his grandmother's terse instruction to mix, Henry mixed. He became a member of the school debating team, where he shone brighter for his command of the facts than for his oratory. He proved himself an organizer when the time came for school plays and other activities. He was well on his way to becoming one of the most popular boys in school, and his grandmother was pleased. She thought that Henry had a "lot of his daddy in him," and she said often that almost everyone liked Henry Rosborough Keplinger because he was a "nice man."

Being a nice man cost Keplinger his job with Santa Fe. He was second in command at the El Dorado installation, reporting only to the chief clerk. Several employees were his subordinates, including one woman. The woman came to him one day at noon and asked for time off to go to her grandmother's funeral. The chief clerk was not in the office at the moment.

Keplinger wavered. Time off to attend a funeral was not unheard of in the 1920s, but neither was it an accepted practice. And both Keplinger and the woman knew that the chief clerk would have turned down her plea in short order. The woman wept, and Keplinger gave in. Soon after the woman left, the chief clerk returned to the office. He reprimanded Keplinger in front of the entire office crew.

White-faced, Keplinger put on his coat and hat and walked out of the office. The chief clerk's apologies the next day were to no avail: Keplinger was through. Nice man though he was, his pride had been cut too deeply for him to return to the office. He set out to find another job and quickly found one, clerking in a refinery.

That evening he told his family they were moving to Russell, northwest of El Dorado and northwest even of their old home in Lyons. He had three sons now. Thomas Earl had been born on April 16, 1918; he was eight years younger than Henry, five years Gilbert's junior.

The refinery job lasted about a year. The installation was shut down when the nearby Fairport oil pool ran dry. But it was a grand year for Henry. He had been downcast when told the family was moving from El Dorado shortly before he commenced his sophomore year. He already had laid plans for work and study. And on his map, Russell appeared to be just a wide place in the road.

He was pleasantly surprised. The high school was a good one, and he was welcomed warmly by both students and faculty. Better still, he found a job that offered more than money for his labor. He went to work in a cavernous general store that supplied the German families in the area with almost everything they needed—groceries, drygoods, hardware.

The proprietor had been reluctant to hire Henry, although he had been impressed by the youth's manners and obvious intelligence. "The customers speak only German," the proprietor had said in his heavily-accented English. "How can you serve them properly?"

"I'll figure out what they want," Henry had replied. He had smiled. "Maybe they'll get some fun out of dealing with me."

It *was* fun for the customers. The women, particularly, enjoyed seeing Henry's frowning concentration break into a smile when he grasped that madam wanted six yards of gingham, not sixteen. On Saturdays the store became as much a social hall as a business enterprise. The men would gather to talk of crops and weather and to drink home brew in chilled cans that once had held tomatoes. They kidded Henry about his "stupidity" and about the pretty girls who boldly flirted with him. They tried to force home brew on him in a friendly fashion, but always the proprietor stopped them. "That boy will drink beer when he decides to," he would say with finality.

Slowly the joking about Henry's "stupidity" ceased. He was learning German with an ease that astonished the farmers and his boss. Within three months, the proprietor claimed, Henry could identify any item in the store by its German name. He did not become fluent, but by the time the refinery closed, leaving his father jobless, he could engage in a halting conversation on almost any subject that arose.

It marked the beginning of his life-long love affair with foreign languages.

□

Fortunately, the closing of the refinery coincided with the end of the school term. Back to El Dorado went the Keplinger family. Keplinger went to work for a Ford dealer. He did well, for he was an excellent bookkeeper. But once again something occurred that stemmed his progress: Henry Ford stopped manufacturing Model T's to make way for the new Model A's. There would be a nine-month gap in production. Keplinger wasn't prepared to sit out the waiting period. He went to work as office manager for the Drane Tank Company. It was a good job, and he was happy again.

Henry, meanwhile, was enjoying his junior year at El Dorado High. He was studying hard, making his usual excellent grades, and involving himself in every school activity for which he could find the time. This was difficult, for he was working after school and on Saturdays at a sandwich shop and selling printing and stationery on the side. And when he could manage it, he would hurry out to Oil Hill in an effort to match his classroom physics against the workings of the field machinery.

The amount of energy compressed in the youth's slender, well-knit body was amazing. "Henry was absolutely tireless," said John Bunyan Adams, Jr., a classmate. "He was always at top speed, it seemed. Even in the classroom it seemed like he was waiting for a bell to ring to send him to something else. Still, he made unbelievably high grades in almost every subject. With it all, the constant work and study, he was what we called a 'regular fellow.' No one ever thought of Henry as a grind."

Adams was Henry's first intimate friend among the scores of warm relationships Henry formed at El Dorado High. And they would share adventures in later years. Ironically, Adams' father was the founder of the Butler County State Bank whose failure resulted in the loss of Henry's $788.32.

☐

Henry was a junior during the 1926–1927 school year. His picture in that year's *El Doradoan* showed a solemn youth with swept-back brown hair, a wide brow, a sharp chin, and wide-spaced, questioning eyes. The caption beneath the photograph noted that he played trumpet in the school orchestra and was on the debating team.

He would have begun his senior year at El Dorado High in September of 1927, destined to graduate in the spring of 1928; but in June, Henry Rosborough Keplinger was transferred to Tulsa, Oklahoma, to set up a Drane Tank Company office in the Kennedy Building.

Henry immediately obtained a summer job with the old Roxana Petroleum Corporation, at that time the Mid-Continent subsidiary of Shell Oil Company. It was a dream job: his duties were light and he had plenty of time to observe all phases of oil exploration and drilling. During the first week he caught the eye of a technologist, Ralph B. Roark, who later would become a Shell vice president.

Roark took Henry under his wing. For three months he tutored Henry while allowing the youth time to tend to his chores.

"He force-fed me," Keplinger would recall many years later.

Henry left his job—and Roark's tutelage—at the end of summer to prepare to enter Tulsa's Central High School as a senior.

In years to come, newspaper accounts of Henry's exploits would report that he was a graduate of El Dorado High School. Many of his closest friends assumed he was a graduate of Tulsa's Central High.

Only a few were aware that he was not a graduate of *any* high school. Late in life he would laughingly describe himself as a "high school dropout."

Asked why he had not corrected the newspaper accounts and the erroneous assumption, he would smile and say, "Too complicated."

CHAPTER TWO

Henry Keplinger stood on the sidewalk in front of Central High School on a cool September morning and stared blankly across the campus. He was in shock. The school registrar had been friendly and sympathetic, but she had been firm. Henry was shy some important credits, she had said. "You can attend classes as a senior, of course, but you can't expect to graduate with your classmates. You'll have to go to summer school to graduate or come back next year to finish up." She had shaken her head to all of Henry's arguments and pleas. He had made his way out of the school in a daze.

Bewildered, hurt, angry, and depressed, he finally trudged to the new family dwelling at 1140 South Gary Place. He poured out his story to Edna. It is likely that her first impulse was to march down to Central High School and state the case for Henry in plain and simple terms. But she said calmly, "You're not going to waste all that time, Henry. You get yourself out to Tulsa University and tell them you want to take the entrance examination. And Henry—you pass it."

Henry got himself out to Tulsa University. He took the entrance examination and he passed it—with the highest total grade, he was told, that had ever been made on the test!

Thus it was that he skipped his senior year in high school and entered the University of Tulsa as a freshman in the fall of 1927. If he ever suffered from the lack of high school credits, it was not evident in his grades during the four years he attended TU or at anytime thereafter when he was working not for grades but for an international reputation in his chosen field.

☐

For Henry Keplinger, the city of Tulsa was his oyster and the university its pearl. The city by 1927 had a population of almost 140,000, having grown by leaps and bounds after 1910 when pipelines had been constructed to carry Oklahoma crude to the Gulf of Mexico and the world beyond. Tulsa had become Mecca for ambitious country boys and girls. Hotels and office buildings sprang up along the wide streets. Railroads ran their lines to Tulsa terminals. Banks serving the oil industry flourished as petroleum companies by the scores established their main offices in the city's attractive towers. Honky-tonks bulged with dancers stomping to what was then called hillbilly music, and the aroma of bathtub gin was as pervasive as the sweet smell of success. It was an exciting city in an exciting era—and Henry Keplinger was seventeen years old.

The first friends he made at TU called him "Kep," and the affectionate diminutive stuck with him throughout his college years. (After college, people began calling him Henry, and he was Henry for the rest of his life.)

"He was interested in everything," a classmate told an interviewer years later about that freshman year. "My God, was he interested! I remember someone saying that it looked like Kep got a running start and just jumped all over the university and the town, too. He was in everything on the campus, it seemed, and all the while he was studying and working at all kinds of jobs to pay his way, and he still found time to prowl around town. We used to laugh and say that Kep went downtown to make sure they were building everything according to his specifications."

In that freshman year, Henry was in the Glee Club and the Pep Club. He was a member of Phi Kappa Delta, the national forensic fraternity, and secretary of Phi Mu Alpha, the national music fraternity. He played in the orchestra and was secretary of the band. He joined the Y.M.C.A., was on the debating team, and worked as a reporter on the *Collegian*.

In the evenings and on weekends he worked in a sandwich shop, and when he found an odd moment he sold printing.

With such a load he managed to earn in the first semester a B- in English, an A in History, a C in Algebra, an A- in Physics, and a *B* in Debate. He improved his grades in the second semester, with a *B-* in English, an *A* in History, a *B-* in Trigonometry, an *A-* in Physics, and an *A* in Bible. In his four years at the University, the freshman grades would be the lowest he would make.

☐

As the freshman term neared its end, Henry received a letter from his good friend John Adams, Jr., in El Dorado. His mother had just bought a new Ford roadster with a rumble seat, Adams wrote. She was going to let him use it to take a summer trip. He planned to visit Washington, D.C., New York City, Boston, Niagara Falls and environs, then cut back across Ontario to Detroit, with a stop at Dearborn to visit the Ford plant. Then he planned to drive west to Marvine Lodge near Meeker, Colorado, where the rest of his family was to spend the summer. Did Henry want to go with him?

The years of work and study had not stilled the wanderlust that had prompted Henry to ride the cow-catcher in Lyons. And he had not idled away a summer since he was seven years old. Still he hesitated; he already had been negotiating to return to a job with Shell, and he wanted the experience as much as the money he would earn.

Edna saw that he was troubled. "Go with him, Henry," she said. "It'll be an experience you'll never forget. And you may learn more about engineering at the Ford plant than you would working in an oilfield."

So he went—and with a light heart.

In that summer of 1928, America was still on Golden Time. The Great Depression was in the near future, but there was not a hint of it on the horizon. So the two young men set out across a land that was bursting with pride and bright promises, driving a brand new car through glorious days and nights toward the beckoning fabled cities.

No matter that they had little money. Bologna and soda crackers and cold root beer made fine traveling fare. Much of the time they slept in the car; in the cities they found inexpensive hotels to bunk in. And a 15-cent Blue Plate Special in a colorful cafe in a strange town was better than a full-course dinner back home.

They took in all the sights in Washington, craned their necks at New York City's skyscrapers, visited all the markers of Boston's storied past. They stared open-mouthed at Niagara Falls, and Henry was as impressed by its boundless energy as by its splendor.

But it was the Ford plant that held the most wonders for him. With a smiling man named Lindsay as their guide, the two youths made the grand tour. By the end of the day Henry had asked a thousand questions, and the weary Mr. Lindsay's smile had all but faded. So Mr. Lindsay was surprised the next morning to see the boys standing at the head of the line of visitors.

"So you're back, I see."

"I think we may have missed a few things," Henry said.

Mr. Lindsay nodded as if he understood, and the tour began.

The next day the youths were back again. If it wasn't too much trouble, Henry told the unbelieving Mr. Lindsay, they would like to watch the tool and die makers, and maybe ask a few questions. Mr. Lindsay sighed. He departed and returned shortly with a middle-aged man in workman's clothing. The middle-aged man was a supervisor of toolmakers, and he spent half the day answering questions. Finally he turned the boys over to Mr. Lindsay. "I think they're going to build their own plant," he said dryly to Lindsay.

The remainder of the trip was anticlimactic. The drive to Colorado was uneventful, and Henry was restless during his short stay at Marvine Lodge. It was too late to get an oilfield job, but he knew he could find work in Tulsa of some kind. But as Edna had said, the trip was an experience he would never forget.

☐

His sophomore year was one of outstanding scholastic achievement. During the two terms he took ten major courses, recording eight *A*'s and two *B*'s. He also was elected president of both the band and the orchestra, secretary and vice president of Phi Mu Alpha, business manager of the *Student Handbook*, and placed on the Y.M.C.A. cabinet. He was obeying his grandmother's admonition to "mix," and he was enjoying it!

He also had been initiated into a fledgling local fraternity, Delta Alpha Delta. It would soon become known as the "Brain House" because so many of the brightest male students in school were members. (Three of them, including Henry, were candidates for the Rhodes Scholarship during the three years he was an active member.) In this band of brothers, Henry would form friendships that would hold firm all of his life. And with these friends, he would learn the taste of bathtub gin. Mitchell Tucker, the fraternity president and years later the publisher of the prestigious *Oil and Gas Journal*, was reputed to be the best gin-maker on campus.

An even more potent drink came out of Organic Chemistry class. There were only six men in the class, including Henry and one of his best pals, Marvin Millard, who would become the Bank of Oklahoma's board chairman. One of the six was laboratory assistant for the class. It was not unusual for the assistant to sneak a beaker of pure alcohol out of the laboratory under his jacket. The six would retire to a nearby hangout where they could buy Cokes. The Coke and alcohol mixture was guaranteed to clean the tartar off one's teeth, according to campus legend.

Henry was a less-than-moderate drinker, however, and would remain so all of his life. He simply took a drink when he wanted one. He enjoyed it, but he had no intention of turning over his mind, even temporarily, to an outside influence.

Despite all of his activity on the campus, Henry had a time-consuming activity off campus that even his closest friends knew little or nothing about: He had put his trumpet to work. With five other musicians, he had formed a dance band—the Tulsa Foot Warmers—and they played almost every night in honky-tonks and lesser-known dancehalls around town. In addition to Henry and his trumpet there was a man on the banjo and guitar, a drummer, a saxophone-clarinetist, a young woman at the piano, and another saxophonist.

(There is a picture of them in that order. The men are wearing tuxedos [Henry had bought one as a freshman to wear while singing with the Glee Club]. The picture is sharp and clear. Henry is instantly recognizable, but not even his closest friends from college days were able to identify the other band members. It may be, as Gilbert Keplinger suggests, that the other members were not college students at all but young musicians Henry had met on one of his off-campus jobs. Gilbert thinks that Henry was not the leader of the band but rather the promoter, the one who got the jobs. It is a reasonable hypothesis, for Henry told an interviewer just months before his death, "I got a lot of jobs because we were cheaper than the others, but still we made quite a bit of money.")

☐

Henry left the band at the end of the school term because he had landed the kind of summer job he had dreamed about—"a kind of assistant to an engineer" with Amerada Oil Corporation. The man who gave him the job was Charles V. Millikan, the company's director of engineering, and Henry would say in later years that he owed as much to Millikan as he did to any man.

He went to work at Seminole, and the very first day on the job two discovery wells came in—the first seismograph discoveries made by Amerada in Oklahoma, in 1929. With two new fields on hand, there was no time for rest and sleeping. "We kept him up for three or four nights," Millikan recalled. "He wasn't accustomed to going without sleep, so we had to awaken him a time or two, and he wasn't easily awakened."

But Millikan was impressed by Henry that summer. "Henry wanted to know everything about everything. He had a thirst for knowledge. And he was ambitious and capable and thorough."

It was during that summer that the "Amerada Bomb," a device used to gauge bottom-hole pressure in a well, was introduced in the oil patch. The bomb, and later variations on it, became one of the most important tools in the oil industry. It was sometimes called the "Millikan bomb," but over the years Millikan was quick to point out that he was not the bomb's inventor. He caused its invention, however, by constantly telling his superiors that such a device was needed and could be designed and constructed.

Finally the job was turned over to an Amerada subsidiary, Geophysical Research Corporation, and a man named Kanenstine, head of Geophysical's laboratory and machine shop, designed and built the tool.

"It was thrilling to us all to receive one, and then to try it and find that it worked just as we thought it would," Millikan would recall. "We got it in January of 1929 and tried it out on several wells at Seminole. Henry wasn't the first to use it, because he didn't come to work until May. But he was on hand when we used it, and he studied it and learned everything about its capabilities."

Millikan was not the insistent tutor Roark of Roxana had been. He simply saw to it that Henry was exposed to as many engineering problems as possible. And he expected Henry to earn every penny he was paid.

☐

In his junior year Henry began a practice that he would continue all of his life. He began "visiting" people. In the workbooks he would faithfully keep during his business years he would note daily that he "visited" with clients and prospective clients. He never met with them or conferred with them or worked with them. He "visited" with them.

As a student he visited the heads of oil companies and engineering firms in Tulsa. "He had absolutely no fear of going right to their offices and announcing he wanted to speak with them," said Mitchel Tucker, his fraternity brother. "He was so assured, so poised, for someone of his age, that secretaries would almost automatically open their bosses' doors for him. Once inside the office, Kep would identify himself and say that he wanted to ask a few questions, that he needed the answers to further his education." Tucker added wryly, "It would have been pretty hard, I imagine, for the top man not to welcome him under those circumstances."

But Henry had more than one string on his bow on these visits. He was letting the top oilmen and engineers in Tulsa know that there was an earnest, intelligent young man at Tulsa University, one who was deeply interested in oil and engineering, and one who would be a graduate in the near future. He was planning for the career he was anticipating.

He followed this theme, also, when he was elected president of the Engineers' Club. The Tulsa *World* reported:

> Henry Keplinger, newly elected president of the Engineers' Club of the University of Tulsa, has announced a series of programs and events that promise to make the

club felt in the professional life of Tulsa, as well as make it a live organization on the university campus.

Consulting engineers and oilmen of the city are giving the club material aid, Keplinger said, in its forthcoming program, which includes a chapel service before the student body, educational meetings of the club, and the annual St. Patrick's dance given by its members in honor of their patron saint. . .

At the special chapel service, March 18, the student body will be addressed by Harry Wright, a consulting geologist of Tulsa, and H. M. Stalcup of the Skelly Oil Co. . .

The second event is to be a meeting of the club March 24 at which K. Winship, vice president of the production department of the Gypsy Oil Co., will speak on 'Production Problems in the Tropics.' Meetings will be held every other Monday evening through the school year. . .

The celebration of the engineering students will culminate in the annual St. Patrick's dance, to be held March 21, in Harwell gymnasium. All engineers in the city will be invited.

A later news story reported that the Tulsa Chapter of the American Association of Engineers had hosted the Engineers' Club at a dinner. "This marks a step forward for the student engineers," the story said, "as it has been their aim to establish amicable relations with practicing engineers of Tulsa. . . ."

Henry started working fewer hours because his summertime oilfield labors had provided him with a financial cushion. And, he was working as a student assistant in the university physics department, a position he would also hold his senior year. Still he found time to be involved in many college activities, including election to the prestigious Student Council; and he was studying German and making excellent grades in the subject, thanks to his work in the general store in Russell, Kansas. He knew that Germany had an abundance of great physics scholars and that great things were being accomplished there in that field.

One of his physics classes that junior year had been on oil and gas separators. National Tank Company of Tulsa was one of the major manufacturers of oil and gas separators. In reviewing the formula National Tank employed in designing a new separator, Henry found a flaw. He wrote the company a nice letter pointing it out.

Within a few days he received a letter from National Tank's president, Jay P. Walker. Henry was correct; a flaw there was, Walker wrote. Would Henry be interested in a summer job with National Tank?

Henry most certainly would!

So he spent the summer of 1930 working for National Tank. Walker was convinced that he had a young genius on his hands and paid Henry accordingly. On his part, Henry did his best to convince Walker that his appraisal was correct. But as always, he learned as much as he contributed.

[ ]

Henry's senior year at TU was so successful, so marked by so many triumphs, that it seemed incongruous that he was denied two things he wanted the most—the presidency of the Student Council and a Rhodes Scholarship to Oxford University in England.

He conducted a spirited campaign for the Student Council presidency, shaking hands and visiting with groups. He printed up flyers and had them distributed. "Keplinger for President of the Council," the flyer proclaimed, " A Student Most Actively Engaged In School Enterprises!" The activities were listed on the inside pages, and at the bottom of the right hand page was printed: "Think it over, are you prejudiced by acquaintances or by acheivement." He had misspelled "achievement," and there were other misspellings as well. Like many engineers and mathematicians, he was inclined to spell phonetically.

Certainly no one had been more actively engaged in school enterprises. Certainly his grades were superior. And certainly he had made many friends. But he lost the election. An old friend suggested that Henry's record may have been *too* good, so much so that it intimidated others less capable and less involved. And he implied that Henry was perhaps too grave of mien. "Kep enjoyed other people's humor, but he didn't generate humor. He wasn't a storyteller by any means, but a listener."

(In later years he would become a superlative storyteller in the family circle, in social groups, and at meetings where he spoke, and the humor which was not obvious in his youth would come bubbling forth.)

Henry was a good dancer. The girls who knew him liked and admired him. He had enough money to "show a girl a good time," as the saying went in those days. But he apparently never got romantically involved with any, and apparently never wanted to do so.

One of his best friends in college, and indeed all through his life, described the first date she had with Henry. She was Anna Ruth (Skip) Watson, a delightful, attractive girl who later married another classmate, Kenneth Hoevel. She had gone to a dance with Henry. When the dance was over and they were walking to Henry's car, Henry said, "I want to show you an oilwell that's being drilled."

She glanced at him obliquely, but she got in the car. She was a popular, well-liked young lady, but she had never before heard a line about watching an oilwell being drilled. So as Henry drove steadily through the night, she wondered.

Finally Henry stopped the car at the side of the road. He got out and went around to open her door. Miss Watson was wearing a long taffeta evening dress, and all she could see out the door was a sea of tall grass. But she got out; she had made up her mind to see this thing through.

Henry lifted a barbed-wire fence so she could duck under it. Then they walked, with Miss Watson trying to hold her gown free of the rough grass. Finally they reached the crest of a rise, and below them the lights on a drilling rig twinkled in the night. They walked on until they were near enough to the derrick to see the driller and the roughnecks at work.

And after a long evening of dance, Skip Watson stood there in the grass for almost an hour while Henry explained the action on the derrick floor.

She said archly to her friends later, "Kep said he was going to take me to see an oilwell, and he showed me one!"

So it may be inferred that Henry Keplinger was not a college

"Romeo." But several years later, when he met the woman who was to be the love of his life, he was an ardent swain at courtship, and he pursued her with a fiery resolution until he won her hand.

[ ]

Some of Henry's friends were quick to find reasons why he was defeated in the Rhodes Scholarship competition, but the reasons lend no credit to Henry, to the winners, to the judges, and to a fine old institution. An incident shortly before the actual competition, however, may have affected Henry's presentation before the judges.

The 14 outstanding male students of the state of Oklahoma, as selected by their colleges, were to compete at the University Club in Oklahoma City. Henry had been chosen to represent the University of Tulsa. Two of the 14 would be selected by the judges to compete with nominees of five other states—Alabama, Mississippi, Louisiana, Texas, and Arkansas.

Henry rode the bus to Oklahoma City. He walked out of the bus station and into a storm of gunfire. Police, pursuing a fleeing robber, shot and mortally wounded the man. He died at Henry's feet, looking up with clouding eyes at the young scholar.

An hour later Henry was in the thick of the scholarship competition. He told no one but his immediate family about the terrifying incident. "He was still upset when he got back home," Gilbert Keplinger said. "It must have bothered him a lot during his presentation."

In any event, he was not selected as an Oklahoma representative. One of the two winners was Carl Albert of the University of Oklahoma, who went on to become a Rhodes Scholar and later a long-time Speaker of the House in the United States Congress.

[ ]

Something fine always seemed to occur to take the sting from Henry's disappointments. In this instance it was the winning of a graduate fellowship in physics at George Washington University and a junior membership in the American Society of

Mechanical Engineers. The Tulsa *World*, in describing the honors, said:

> Membership in the national engineers' organization came to Keplinger with the announcement that his paper, 'Stresses and Strains in Unfired Pressure Vessels,' delivered in competition before the annual meeting of the Mid-Continent section of the society, had been awarded first place. Roy White, national president, New York City, was the honor guest at the meeting.
>
> In achieving this distinction, Keplinger presented his paper in competition with students from Oklahoma University, Oklahoma A.&M. College, and Arkansas University. It will be published later in the *Oil and Gas Journal*.
>
> The student, who has received many honors during his college years, plans now to accept the physics fellowship at George Washington and continue his research there next fall. . . .

Much of the information that had gone into the writing of the paper had been gathered during his summer work with National Tank Company. Now he began work on another paper, "Physical Problems in the Separation of Oil and Gas," also based on knowledge he had gained that fruitful summer. It was presented before the Oklahoma Academy of Science, into which he was initiated.

But before he had written the paper he had written a dozen letters to various individuals and companies from coast to coast, asking for information. The letters were a fine blend of humility and youthful assertiveness. He was humble in asking for information, but at the same time he was making it clear that he and his fellow engineering students at TU were not dullards. A typical paragraph read: "My little experience with the separation of oil and gas leads me to believe that the surface has only been scratched toward finding critical data relative to separation characteristics. We (the class) have learned that certain phenomena occur under given conditions, but as yet we are unable to forecast with precision the results on projected problems. Our knowledge covering velocities of a rising column of gas, through which various size droplets of oil of different characteristics will fall, is limited. . . ."

He received a surprising number of replies, full of information and good wishes for his future.

While he appreciated the information, Henry wanted more than good wishes. The stock market crash of 1929 had been followed by the Great Depression, and now, in 1931, America had almost been brought to its knees economically. Many of Henry's classmates had been forced to leave school. Henry was fortunate in that he lived at home and that his father was still employed. And he had worked hard and husbanded his resources wisely. But without the fellowship at George Washington, the future would have looked much grimmer. The fellowship was months away, however. Directly ahead of him was a bleak summer.

So more and more letters went out—this time containing copies of his paper on which he asked for comments. In almost every letter he wrote: "The paragraph in the paper entitled 'Results of Separation' is probably in error for the most part. What should have been said is that the oil is spread out in a thin film and the gas evaporates easily, any mechanical entrainment of the higher hydrocarbons is removed by centrifuging, impact, and impingement. . . ."

Thus he invited comments. And he got them. But he got no job offers. In one instance he carried on a great deal of correspondence with the Trumble Gas Trap Company of Los Angeles. In one letter, he wrote:

> There is one thing about separators, your separator, I do not see and that is the inside float. How can it work under great pressures and oil and gas volumes? Would like some facts and examples which show it to be better than outside float, or at least equal in efficiency and service. The only advantage of the inside float is the lower cost. I would greatly appreciate your help on the above problem.
>
> If you decide to enter the Mid-Continent I know we could get a chance to try out one of your series installations at once. If the test proved successful, we would be a success in the high pressure trap business. An efficient service organization would have to be maintained and there would also be other expenses. If I knew the exact cost of fabrication of the separator and the valves, I could estimate your profit and tell you whether it would be ad-

visable to enter the Mid-Continent. If you can see a clear way to make a profit, I know we could sell the separators. . . .

If Trumble Gas Trap Corporation ever entered the Mid-Continent, it was without Henry as its agent. He never received the kind of answers from the company he was so badly wanting. Companies simply were not in an expanding or hiring mood.

But even the Depression could not dim Varsity Night, the annual all-school show, and Henry was general chairman of the celebration. The show featured all manner of contests between the fraternities, sororities and non-fraternal groups, and a play, "Campus Nights."

The *Varsity Night Review*, published by the Student Council, reported that Henry was in charge of making all the arrangements for the annual celebration and had worked out all the details for the evening's entertainment. Henry was quoted as follows:

> Traditions and jolly good times spent together in university life remain forever in our memory as golden treasures. In after years, when life has grown somewhat dim and gray, we will recall our joyful, carefree years spent in college life, and the memories are bound to quicken our step and make life more pleasant. One of the most delightful nights that will be reminisced will be Varsity Night . . . it will be bigger and better than any past performance.
>
> Varsity Night welds the entire student body into a compact organization for promotion of school spirit and loyalty to school enterprises. It promotes extensive cooperation among the students and among student organizations. To succeed in the great endeavor, it will take the individual support of every student and faculty member in the university. For Varsity Night to be remembered, it must be a grand and glorious success, and I appeal to every student to enter into the project with great zest, and Varsity Night will live forever in our lives. . . .

Varsity Night, let it be noted, was a grand and glorious success!

☐

Graduation time at the University of Tulsa was all of the holidays wrapped into one for Edna Keplinger and Nora Suffield. They saw their pride and joy graduated Magna Cum Laude. They saw him and only seven others initiated into Phi Gamma Kappa, the honorary scholastic fraternity, for maintaining A-averages through four years of college work. And they saw him honored with the coveted Howard Acher gold medal, awarded annually to the school's outstanding graduate. The faculty committee making the selection considered character, scholarship, leadership, personality, and campus and school activities. The gold medal proved to Nora Suffield that her grandson had "mixed" as she had told him to do when she gave him the shining trumpet.

Some of Henry's pals in Delta Alpha Delta regretted his departure. He had been their advocate when members got in trouble with the administration for their high-jinks and shenanigans. "Kep would always go to the authorities and talk for the guys," Mitchell Tucker recalled. "The authorities simply couldn't believe the boys were bad if Kep associated with them, and they always decided to forget the incidents."

And there was something else. In 1981 an interviewer called on Alvin DuVall of Houston, who had retired after a long and distinguished career in journalism. DuVall had been a member of Delta Alpha Delta back in those days.

"Sure I knew Henry Keplinger," DuVall said. "Knew him well." But then he quickly amended his statement. "I mean I *remember* him well. I'm not sure anyone knew Henry well. Everyone knew him. He was liked and respected. But he was a private person."

Why did DuVall remember Henry so well, the interviewer asked.

"Because of something he did," DuVall said. "Something I could never forget. It was in the Depression. My mother died, and there was total confusion at my home there in Tulsa. I was standing in my front yard and people were coming and going, old family friends, when Henry drove up and parked at the curb. I knew him, but not well, and I was surprised at seeing him.

"We met in the yard. He talked with me only a few minutes, less than five, I'm sure. I can't remember a word he said. But whatever he said did me a world of good. It heartened me. And I've never forgotten it."

Throughout his lifetime, hundreds of persons would remember Henry Keplinger for similar thoughtful acts. It was a characteristic not instilled in him by his parents or his long-widowed grandmother, but granted him by the God he acknowledged in his daily prayers.

□

There had been a symbiotic relationship between the University of Tulsa and the city's business and civic leaders from the school's beginning. The city looked to the school to supply it with future leaders. The university looked to the city to keep its students employed so that they could graduate and become future leaders. And the city's leaders had been most generous with their contributions to the university's growth and general welfare. It was unstated but generally accepted that future leaders would continue this practice.

The Depression made the relationship shaky; there were not enough jobs to take care of the existing labor force, much less recent TU graduates. Not even honor graduates.

Henry had the fellowship awaiting him at George Washington University with free tuition and a $600 stipend. But that was three months in the future.

Then he got a telephone call from Jay P. Walker of National Tank. Did Henry want to go to East Texas for the summer? Walker asked. It looked like East Texas was going to be a big field, and National Tank was having trouble with its separators there. He had a job if Henry wanted it.

Henry wanted it.

CHAPTER THREE

C olumbus Marion (Dad) Joiner discovered the great East Texas oilfield on October 5, 1930, with the third well he drilled on the Daisy Bradford farm near Henderson in Rusk County. On December 28 the Lou Della Crim 1 roared in near Kilgore, and on January 26, 1931, the Lathrop 1 extended the field even farther north into Gregg County. After that, derricks marched in every direction, studding the landscape in five counties. Oil was spouting from a reservoir 45 miles long in a north-south direction and 5 to 12 miles wide. In all, it covered more than 140,000 acres. And billions—not millions—of barrels of oil lay beneath those acres in a blanket of Woodbine sand.

While factories stood idle across the land, and heartsick, hopeless Americans stood in bread lines from New York to Los Angeles, East Texas, in the spring of 1931, was caught up in the most frenzied boom in the nation's history. Kilgore had been a hamlet of 700 souls when the Lou Della Crim 1 was brought in. Before the week was out, 10,000 boomers had invaded the area. Kilgore was a madhouse.

And it was to Kilgore that National Tank sent Henry Keplinger. Men were sleeping in tents, in cars, in wagons, and on the ground itself, but Henry's first letter to Cecil Wells, National's secretary-treasurer, declared: "Have a mighty fine place to stay in Kilgore and like everything fine. . . ." He had been befriended by a driller, Jim (Doc) Hurley, who had a bed in a crowded rooming house. Since Hurley worked at night and Henry worked days, they took turns using the bed. There were no weekend problems: everybody in the East Texas field worked seven days a week.

In addition to being an engineer and serviceman for National Tank, Henry also was a salesman—and National Tank had a lot to sell. Along with the oil and gas separators, the company was marketing welded flow tanks, casinghead storage tanks, absorber and scrubber towers, pressure and vacuum tanks, bulk station gasoline tanks, bolted and wooden tanks, pressure and vacuum valves, liquid level regulators, and plates, sheets, angles, channels, and tank bolts.

And the competition was fierce. Further, almost every manufacturing and distributing company operating in the field was learning that equipment that functioned perfectly in other fields seemed to be attacked by gremlins in East Texas. Henry kept up a steady flow of correspondence with the National Tank office in which he described mechanical problems and recommended solutions.

He was having trouble, for example, with the semi-balanced gas valve on National's separators. He wrote:

> It seems that the valve is not sensitive enough for a condition of large oil volume and small gas volume. It may be possible that only a four-inch gas valve should be used for all separators. Gas volumes never run over seven or eight million cubic feet, that is per day. . . .
>
> Some think that the top seat should be built a little larger and the guides on the valves should be smooth instead of rough. A grinding noise may be heard when the valve is worked up and down. There is also the possibility of dirt and sand collecting between the guides and the body. There may be considerable friction in the working of the stem in the stuffing box.
>
> The latest trouble with the gas valve arose on the Golding-Murchison, Thompson Lease. A No. 7F National separator is handling two wells. At any rate of production, large or small, the National gas valve would not hold a constant pressure on the separator. The pressure at the inlet side of the valve would build up to 32 lbs/sq. in. and release suddenly to 18 lbs./sq. in., then gradually build up to 32 lbs./sq. in.; the oil volume was approximately 80 bbls. an hour. When the pressure dropped, the gas vent line would throw out oil. This condition happened at all separator pressures, 18 lbs./sq. in. to 100 lbs./sq. in.
>
> The producing company put in a new Smith valve and

the separator gave good separation and did not throw out oil from the gas vent line. The valve operated at a constant pressure. It appears that poor separation was due to siphoning action caused by sudden release of pressure. If you want any information I will get it immediately. Will send in valve if you want it.

Golding-Murchison purchased two more separators, but did not buy the gas valves. They plan to use the Smith valves.

I would like to know the serial numbers on the six small separators you were building about the first of June. The separators contained a long inside skirt with a three louver opening in the top.

I should like to know the formula you used for figuring gas line capacities, and also some mercury and the calibrated pressure gauge.

Some of the operators have asked why we do not have gasket material for the oil and gas valve flanges. At times the service men go and get gasket material. At times the lever collars and the ends of turnbuckles need considerable filing to remove paint and burrs. The turnbuckles are the right length now. The diaphragm control will be of great assistance in combating the paraffin problem this coming winter. . . .

This letter was written when he had been in East Texas only a few weeks. By August he had lost his diffidence and was writing with certitude; a man learned quickly in coping with the eccentricities of the Black Giant or he was soon on his way to other climes. "I have the special gas valve ready to go," he wrote the home office in August. "Houston Oil Company will install the valve on one of their present D.C. separators, out on the vent line, just past the riser, and then plug the riser. Will lay a short line away from the valve. Waiting for wells to open." The "special valve" was the result of Henry's studies. He had sent his observations back to National Tank and the company had built the valve to overcome the problems Henry had been describing. It worked well on the Houston Oil Company separator, and soon was in demand throughout the field. "Mr. Harmon is going to install his own baffles in the old Smith separator, *but we will get the valve order,*" he wrote triumphantly.

With his work going better, Henry had time to enlarge the

scope of his studies of the great field. In one of the letters he wrote to his old mentor, C. V. Millikan of Amerada, he said: "I will tell you what I believe. The gas is completely dissolved in the oil at the prevailing bottom-hole pressure and temperature. There is no free gas in the sand, and the contents of the oil reservoir are liquids. *The oil is not moved through the sand by gas, but by hydrostatic pressure.*" [Editor's note: That last sentence was italicized by the author, not Henry, and for the following reason.]

As soon as it had become obvious that the field would be a giant one, oil company scientists began trying to determine what kind of energy forced the oil through the borehole to the surface in such volume and with such power. Until the East Texas field was discovered, scientists had concentrated their studies almost exclusively on the role gas (both free gas and gas in solution) played as a reservoir energy source. But there was no free gas in the East Texas field, as Henry had noted in his letter to Millikan. And Ben Lindsly of the Bureau of Mines determined that East Texas crude was undersaturated with gas, and thus eliminated gas in solution as the energy source.

So the scientists turned their searchlights on water. There was salt water beneath the Woodbine and possibly within it, but such water would be static, incapable of exerting pressure on the oil, unless it was replenished by surface waters.

The Woodbine outcropped in the Dallas-Fort Worth area, some 150 miles west of the field. So studies were made to see if water entering at the outcropping could flow through the Woodbine 150 miles to displace crude produced from the field. Cores were taken and studied for porosity. It was decided that water could not trickle through the rocks fast enough to displace the oil being produced.

Humble Oil and Refining Company had a brilliant research team working on the puzzle—T. V. Moore, H. D. Wilde, R. J. Schilthuis, and William Hurst. Henry had met Moore and had talked with him several times. On one occasion Henry suggested that the team might want to consider the compressibility of water in finding an answer.

The scientists knew that water was compressible, but so infinitesimally that no one interested in reservoir energy had ever mentioned it before. But Moore and his teammates started working on the theory that expansion of water in the reservoir was the energy source.

The team's conclusion was widely publicized in the field: "The data on the East Texas field indicate clearly that water drive or water encroachment is by far the most important agency in maintaining the reservoir pressure or in producing the oil. It appears that the water moves into the field by virtue of its own expansion upon reduction of pressure, and not, as in many cases, by flowing through the entire formation from its surface outcrop to the field. . . ."

Henry mentioned his conversation with Moore, perhaps for the first time, in a brief interview shortly before his death. It was several months after his death that a researcher found a carbon copy of the letter to Millikan among his papers. There is no record of his ever attempting to take credit of any kind for the important finding.

Not everyone in the East Texas field believed that the Humble men had arrived at the correct conclusion. Many operators simply didn't want to believe it, or admit they believed it. For what the Humble men were saying in effect was this: when you take oil from the reservoir in a proper fashion, it is replaced by water, barrel for barrel, pressure is maintained, and a greater amount of crude can be recovered from the reservoir. If you produce recklessly and unevenly, water will flood the Woodbine, oil will be lost beyond recovery, and the reservoir's pressure will be dissipated.

E. O. Buck, a brilliant state engineer, proved the Humble team correct on December 17, 1932, more than a year after Henry had left the field. Using Amerada bombs, Buck obtained the bottom-hole pressures of 28 wells in the field, starting in the west where the water drive originated, and working east to where the Woodbine pinched out against the Sabine Uplift.

Wells on the extreme west side of the field registered bottom-hole pressures of about 1,400 pounds per square inch. Wells on the east flank recorded pressures of about 700 pounds per square inch. Disturbingly, the Joiner well, the first in the field, also on the extreme east flank, had shown more than 1,600 pounds per square inch of pressure about two years earlier.

The field was shut down for four days. Then Buck tested his

28 wells again. There had been little pressure change in wells on the west flank, but the 700-pound pressure in the wells on the east flank had risen to 1,300 pounds.

This was proof that the drilling and producing orgy was destroying the reservoir, and was the first step toward implementing sane conservation practices.

Henry would say many years later that his few months of work in the East Texas field, when he was 21 years old, was one of the great "learning times" of his life.

☐

At George Washington University Henry taught physics to aspiring medical doctors to earn his tuition and the $600 stipend. And he extended the scope of his studies to nuclear physics and to his then paramount interest, geophysics. He saw seismology as the complement to the engineering, geology, and petroleum engineering courses he had studied at Tulsa University.

It will be recalled that he had worked for Amerada in the summer of 1929 when that company made its first seismic discoveries in Oklahoma. His interest in the subject was heightened in East Texas where every geophysical device known to science had failed to find a hint of the vast wealth beneath the soil. When Dad Joiner found the field by the seat of his pants, seismologists and geologists, who had condemned the area as barren, were jeered in oil camps around the world. Later it was learned that the instruments had found none of the structural indications of oil because there were none. The vast field was a stratigraphic trap without a single salt dome or fault system within its perimeter.

But oil finders nevertheless were confident that the torsion balance, the refraction seismograph, and later the reflection seismograph could find oil because they had previously found it in abundance. The first seismic instrument used in Texas and Oklahoma was the torsion balance, in 1922. It was used for measuring the pull of gravity in the hope that it would help locate salt domes and other formations having a density differing from that of surrounding rocks.

In 1924 the Marland Oil Company began using the refraction

seismograph. It was expected to help determine the depth and character of formations by recording the speed and quality of shock waves sent through them and refracted to the surface. Soon both the refraction tool and the torsion balance were in widespread use.

Some of the best and largest fields of the period were drilled as a result of seismic testing. In Sugarland, near Houston, Humble Oil and Refining Company geophysicists found a salt dome buried some 3,300 feet below the surface, and the drill found oil.

The prototypes of these devices were conceived in Germany, and a Dr. Minthrope of Gottingen University had spent some time in the Gulf Coast area as a consultant on their use. Henry was elated when he read in a Washington newspaper that Minthrope was to be at an upcoming geophysical convention in Houston. He immediately made plans to attend.

In his customary manner, he managed to meet Minthrope and have a long conversation with him. Minthrope was pleased that Henry spoke German fairly well as a result of his work in the general store in Russell and two semesters of study at Tulsa University. In turn, Henry found Minthrope brilliant and fascinating.

Minthrope told Henry that at Gottingen University a predecessor had devoted his life to studying earthquakes and determining the extent of earth disturbances with seismic devices and mathematical computations. His formulas for locating an earthquake's epicenter were regarded as standards.

Minthrope much later decided to expand on the other man's work by testing to see if earthwaves could reveal geological structures. He had built a high tower from which was dropped a heavy iron ball—heavy enough to disturb the earth and allow Minthrope to collate dependable and verifiable earthwave data. That had been the primitive beginning, but it had been responsible for the birth of the seismic tools now in use.

Minthrope suggested that Henry spend his summer in Germany, attending classes at Gottingen University and making excursions to other schools and locations where he could broaden his knowledge. Henry's imagination had been fired by Minthrope's talk. He totaled up his finances in his mind and told Minthrope he would see him in the summer.

☐

On his twenty-second birthday he sailed for Europe aboard the *Leviathan*. At this time of his life he was experiencing the best of possible worlds, satisfying at the same time his thirst for knowledge and for travel. The Great Depression was now worldwide, and the American dollars in his pocket had depreciated in value far less than European currencies. If he lived carefully, he could live like a prince. And he knew that a second term of his fellowship at George Washington University—with its paid tuition and $600 stipend—was awaiting him in September.

He had matured. Photos of that time show that his face and body had filled out, though he was still trim. The once oil-sleek hair was now loose and wavy. He was an attractive, well-dressed young man whose work and study habits had toughened him mentally and physically without diluting his basic sensitivity. The Charles Henry Keplinger walking the deck of the *Leviathan* already possessed the essential characteristics of his adulthood, and nothing would ever alter them.

From London he sent home dozens of postcards and pictures. One can only guess how he must have felt when he mailed one postcard to his brother, Gilbert. The picture was of Tom Tower and Quadrangle, Christ Church, Oxford. The card said: "Visited in Oxford all day, 53 miles from London, came by bus. Wonderful, with buildings and works of art that date back to 14th century and older. Went down to banks of Charles River and saw the games and lots of canoes. . . ."

If he left Oxford and London with any sadness, he apparently soon shed it. The flood of postcards continued from Paris, The Hague, and other stops on his way to Gottingen University.

Gottingen was an ancient city of about 45,000, some 67 miles from Hannover in the beautiful valley of the Leine River, dotted with magnificent old churches and public buildings that were constructed with loving hands in the fourteenth century. But its latter-day fame rested on its university, which opened its doors to students in 1737. It became a literary center which attracted scholars from all over Europe.

By the time Keplinger arrived in 1932, the school had become a scientific center in a city noted for the manufacturing of scientific, optical, and surgical instruments. And now the students came from all points of the compass.

Keplinger immediately plunged into his studies, devoting most of his time to work he would continue at George Washington University for his thesis. In later years he would say that he "developed a steam attack and determined the heat rising from the core of the earth . . . a method whereby we measure the heat conductivity and the heat loss over parts of the earth. . . ."

He spent no more than a month on the Gottingen campus. His research took him to learning centers at Munich, Dresden, and Berlin, and he made side trips to other German cities and to Copenhagen as well. From the spate of postcards and pictures, it appeared that he was determined to visit every castle, church, and other historic landmark on the continent.

In most of the cities, he could stay in a middle-class hotel with breakfast delivered to his room for only 45 cents a day. He sent home cards showing these places, with arrows pointing to his rooms. In Berlin, he wrote Nora Suffield that he had ridden the train from Gottingen for only $4. He continued:

> A big celebration in town today. Germany on Aug. 11, 1919, became a republic. Tuesday I visited Berlin University Physics Institute and walked around on Unter den Linden, a famous Strasse which has hundreds of nice buildings on it, is 250' wide.
>
> Yesterday I visited the American consul and a friend in Agri. office. He invited me for a trip Sunday. The rest of the day was spent at the Geological Institute and will go back there this morning. (My landlady just brought my breakfast in.) Tuesday night I went with a friend to visit his sweetheart. She has a sister and so I had a good time. Last night I went to talking movies. . . .

He made friends easily wherever he went. His command of the language, shaky though it was in the beginning, became strong enough for him to be at ease in and out of the classroom. He sent home pictures of his friends at school, including photos of his "best American friend," Herman Eilers, dark-haired, square-jawed, handsome as a movie star. There were pictures of Keplinger quaffing beer while seated on a freight wagon, waiting for a train. He wore a gray felt hat and a gray suit, and his camera was tucked under his arm. And there was a picture of him in the same suit, wearing a Tyrolean cap, and standing beside his trusty bicycle.

He mentioned this bicycle in a letter he wrote to the editors of the Journal *Mining and Metallurgy*, in which he outlined some of the highlights of his stay in Germany:

The widespread use of electricity for power is especially noticeable, it being much preferred to steam or oil engines. However, street gas lamps are frequently used, as some of the cities have large investments in gas plants. Freiburg, for instance, has 1,200 gas street lamps, automatically regulated by diaphragm spring valves. At midnight, the gas line pressure is increased 5 lb. for 5 min., which operates a catch arrangement on the valve stem, and thereby reduces the gas volume by one-half. At daylight, the lamps are turned on similarly.

In railroad construction, the rails are longer than in the U.S., about 60 ft., and the junctions between the rails are opposite each other instead of being staggered. In many of the improved roadbeds steel ties are used instead of wood. They have proved satisfactory after being in use a long time, though they make more noise, as is easily perceptible when one is riding in a coach.

My visit to the Hannover oil fields was particularly interesting. For this purpose I found it desirable, from the standpoints of both convenience and economy, to purchase a bicycle. It was not exactly a mechanical marvel, costing only $10 new, but served the purpose and I sold it after six weeks's use for $5. Bicycles are much more common in Germany. I visited the two important fields of the country, Wietze, which is about 32 km. northeast of Hanover, and Nienhagan, which is about 20 km. southeast of the Wietze field.

Wietze, the oldest field, produced its first oil in 1859. Here the old production methods may be observed. Bailing and pumping are used to bring the oil to the surface. Every power need is met by electricity. The quantity of gas available is not sufficient for gas engines.

An oil mine proved of great interest. A 300-ft. shaft has been sunk into a shallow depleted oil sand which contains little gas. Drifts are made in both directions at the bottom of the shaft. The sand is mined and carried by a continuous conveyor to the surface where it is washed with hot water in cylindrical tanks with a cone-shaped bottom. The process is continuous; sand enters the top and flows out at the bottom, while oil and water flow over the top. This gives a clean, shiny sand that contains only minute quantities of oil. The sand is returned underground after washing. The oil and water must be chemically treated before the oil is run to the pipe line or loading rack. Over 600 bbl. of oil a day has been recovered at the mine. Today the production is a little less than half that, but can easily be increased to 600.

Nienhagan is the most modern field and compares with our modern fields in every way except that the 70-foot derricks are completely covered with boards to keep out visitors and conceal operations. The production, engineering and geological departments have their offices in the center of the field beside a fine machine shop which makes, besides other things, all the drilling bits used. All the operations are closely checked and splendid cooperation was found.

Electricity is exclusively used although they have plenty of gas which can be utilized. A modern electric rotary rig with 122-ft. derrick was in operation in a deep test which was to be drilled about 2,500 ft. The 1,800 to 2,400 ft. wells are equipped with electrically geared pumping jacks and are using both Axelson pumps and those of German make. A gas recovery line covers the entire field and feeds into a modern charcoal adsorption gasoline plant. Every well is connected by a narrrow-gauge railroad which leads from the machine shop and affords the only means of moving equipment.

My visits to the wells were made on bicycle and afoot. In the Wietze field, the field superintendent used a motorcycle and the field engineer in Nienhagan had a bicycle. Both regions remind one of the East Texas area, being rather sandy, not quite so rolling, but completely covered with pines which were planted by the government. The tops of the derricks can just be seen over the trees. . . .

□

He sailed home in early September on the *Aquitania*. He returned to George Washington University to find his long-time friend John Bunyan Adams, Jr., in attendance. Adams had transferred from the University of Arizona. They tramped together around the nation's capital, seeing the sights they had seen together on their memorable pre-Depression vacation. They took canoe trips on the Potomac. There were four other men from El Dorado at the school, and Keplinger sent home pictures of them all.

The school year passed. He completed his studies and received his Master of Arts degree on June 7, 1933—his twenty-third birthday.

In a tragic period in American history, when almost as many students fell by the wayside as completed their studies, Keplinger, by dint of grueling work inside and outside the classroom, was now, in an expression common to the time, "educated to his eyeballs."

But in 1933 the Depression still held its grip on the American economy and spirit. So Keplinger stepped off the campus into a cold world where one man's future looked as bleak as another's, education notwithstanding. A man of intelligence and imagination could hope, but the promise of America seemed lost forever.

CHAPTER FOUR

H e couldn't find a job in Tulsa. He went from one oil company office to the next, but there were no openings. Company officials were aware of his background and training, but they were pressed hard to keep their veteran employees on the payroll. A dozen times he heard the phrase, "If something opens up. . . ." He had held high hopes of going to work for Shell, but Shell had nothing for him. "We want you, Henry," Shell officials told him, "but we can't take you on right now. We'll grab you when the time comes."

In later years he told an interviewer, "Some friends finally took pity on me and got me a job roughnecking." The friends were some men he had met earlier who worked for Loffland Drilling Company, and they got Keplinger a job on a drilling crew in West Texas, near Big Lake.

The work was hard, the weather was hot and the landscape empty, but Keplinger liked it. He enjoyed the rough camaraderie on the derrick floor. And he was sustained by a firm belief that a call would come from Shell. So he labored with a light heart, learning all he could, taking with a smile the friendly ribbing of the driller and the more experienced roughnecks.

Five months passed before his hope was realized. Shell *did* call—and Keplinger returned to Tulsa as a trainee engineer. His first job was at his alma mater. TU had provided Shell with laboratory space for determining porosity and permeability of core samples from all over the Mid-Continent. Keplinger was placed in charge of the installation. He described his work there years later: " . . . the basic research on core analyses was sent to six laboratories over the United States. We had these laboratories check my fundamental principles of rapid determination of porosity formations, the permeability in relation to both source material for accumulation of oil and as reservoir rocks which would entrap oil in the future. . . ."

He was plucked out of the laboratory by Paul Guarin, who had been instrumental in hiring Keplinger the summer before he began his college career. Guarin was now field superintendent for Shell at Lucien field in north central Oklahoma. Lucien field had been unitized—that is, it was produced as a unit— with Shell as operator for all who held acreage in the field. At that time only two other fields in the U. S. had been unitized, Kettleman Hills in California and Van field in East Texas.

It was at Lucien field, then, early in his career, that Keplinger learned the blessings of unitization. Such a program, he was taught by Paul Guarin, could mean the end of hasty drilling and overproduction which had in the past and would in the future cause untold damage to fragile reservoirs. He would become one of unitization's staunchest advocates later in his career, and testify in countless unitization hearings.

He left Lucien field and Guarin's tutelage as a full-fledged engineer. He would return some time later as district engineer for Lucien field and later be placed in charge of its production. But now he headed for California. As he testified at a court hearing years later: "I spent a year in Los Angeles and surrounding areas, working on geological problems, along with engineering. I was quite young then, but I did considerable work with the complicated Ventura structure. I spent quite a bit of time at Long Beach and Bakersfield, and I was very familiar with the geology at Kettleman Hills and the entire San Joaquin Valley. . . . I was assigned to the geological department, surface geology, in Ventura, the whole area from the surface standpoint, and I tramped many parts of the area, viewed the outcrop, so my business has been both geological and petroleum engineering. They interrelate because the parameters of a reservoir is dependent upon the geology. If you miss the parameters, then your petroleum engineering has no meaning. . . ."

With the California experience behind him, he made a stop

in Texas to work with Shell engineers, then returned to his beloved Tulsa as chief appraisal engineer for Shell's Mid-Continent Division.

And in the summer of 1935 he took the second vacation of
his life. He borrowed his father's 1933 Chevrolet, gathered up
his brothers, Gilbert and Tom, and headed for Mexico City via
the Pan-American Highway, which at that time was still under
construction. Gilbert had just been graduated from Oklahoma
A & M, and Tom was still in high school.

It was an adventure none of them would ever forget. Time
after time the road would peter out, and Keplinger would steer
the faithful Chevrolet over prairies and down gulleys until the
road reappeared. At other times they would be held up by
workmen who saw no reason for the trio to be riding along
their handiwork so soon. Keplinger talked his way around
them, occasionally doling out a peso or two.

He bought gasoline wherever he could find it, sometimes
walking back from a farmhouse to the stalled car with the pre-
cious fluid in a rusty bucket. A few times he bought it from
workmen who had blocked his passage. But they made it to
Mexico City and reveled in its strange sights and sounds.

The way back was no easier, and when they finally reached
Tulsa, the car was ruined. Keplinger bought his father a new
1935 Chevrolet.

□

It was Thanksgiving night, 1935, and Henry Keplinger, 25
and still unattached, was standing in the lobby of the Wells
Hotel in Tulsa. Some friends had invited him to a cocktail party
in the hotel, and he had paused beside the elevators.

At that moment an attractive young woman entered the hotel lobby, and Keplinger's life was changed forever. She was
wearing a bright green coat with a lynx collar, and a tiny green
hat sat saucily on her warm auburn hair. A young man was
with her, but Keplinger hardly noticed him. As the woman
crossed the lobby, Keplinger saw that her mobile face was alive
with interest for anything that crossed her field of vision. It
was a charming face, a bit elfin, and to Keplinger the loveliest
he had ever seen. An emotion new to him gripped him, and he
told himself, "There's the girl I'm going to marry!"

And, luck of luck, the woman and her escort had been invited to the cocktail party! With an eagerness he had never before experienced, Keplinger introduced himself to the smiling stranger.

He learned that she was Louise Spang of Butler, Pennsylvania, that her father was in Tulsa on business, that she was visiting with her aunt, Mrs. W. R. Walker, that her escort was a cousin, Artie Walker, and that she was going to a dance with Walker later in the evening. It would have been obvious to anyone who cared to observe the couple that she was as interested in Keplinger as he was interested in her. He asked her to go to a dance with him later in the week, and she said she would.

On that first date they had drinks with some of Keplinger's friends before they went to the dance at the University Club. Keplinger simply was different from the young men Louise had met in Butler, at Sullins College in Bristol, Virginia, and at Muskingum College in New Concord, Ohio. He was poised, full of cheerful confidence, and most attentive. She was happy in his company.

When she said she was interested in Indians and Indian lore, Keplinger took her to Pawhuska, seat of the Osage Tribal Council, in adjoining Osage County.

When she said she had never been to a night club, he took her to one that featured an exotic dancer. They had hardly been seated when suddenly the band stopped playing in mid-tune. Dancers froze. Silence.

Henry leaned across the table. "This place is being shaken down," he whispered. Louise turned to face the door. Two men in trenchcoats with their hands thrust ominously in their pockets had entered. They began striding toward the bar where the night club owners stood.

"They're going to shake them down for protection money," Keplinger said. He grabbed Louise by the arm and hurried her to the cloakroom. He got their coats. By now the thugs were pistol-whipping the owners while the rest of the patrons stared in horror.

Keplinger got Louise out of the club and they rushed to the car on a prairie parking lot. Before they could drive off, they saw the thugs propping up the owners on a rail fence to shoot them.

On the ride home Louise kept telling herself how brave Keplinger had been, so quick to act, in getting her out of the night club. The next morning the newspaper was full of news about the bloody doings at the club. Her uncle, Will Walker, showed her the newspaper. "Isn't this where you and Henry went last night?" he asked. She said that it was. Keplinger came by to see her that evening, and again she thought of how brave and resourceful he had been.

Said Keplinger to his love: "I was scared stiff because I had a Shell company car."

☐

Louise Spang told Keplinger that she had a boyfriend back east, but she was so enjoying being courted by two men she decided not to tell Keplinger that she was practically engaged to his rival. She was only 20, and it was wonderful to have two beaus vying for her affection. "I'm pretty serious about him," was all she would say about the distant boyfriend.

Finally her father completed his business in Tulsa, and Louise returned to Butler with him, leaving the disconsolate Keplinger to pursue his courtship by long distance telephone. (Louise didn't want her three sisters, Margaret, Lillian, and Charlotte, to have an opportunity to intercept his letters, so she told Keplinger not to write. However, the sisters couldn't help but notice that she talked long and often on the phone with the man from Tulsa.)

The Spang family was a Pennsylvania institution. Louise's grandfather, George Ashbell Spang, had died almost a year before Louise met Keplinger in Tulsa. With the help of his three sons, Ferdinand J., Loyal B., and C. Everette Spang, he had built a large oil tool manufacturing company that was headquartered in Butler and had offices in almost every oil-producing state.

In its report on the death of George Ashbell Spang the Butler *Eagle* noted:

> One of the outstanding accomplishments in his career as a manufacturer was the development of the method of making drilling and fishing jars from a single bar of steel. Formerly jars were made by fire and hammer, welding four pieces of steel. The welding caused a weakness which

in use frequently resulted in costly fishing jobs. Mr. Spang's patent did away with these welds, consequently a stronger piece of steel could be used which was heat treated, an operation prohibited in the welded type. . . .

Always a leader and a progressive manufacturer, he continued in the manufacture of the cable system of drilling and fishing tools, possessing many patents covering such tools. . . .

F. J. Spang, his oldest son [and Louise's father], has been plant manager and in charge of all manufacturing for the past ten years, and with his father designed and perfected many improvements and inventions on the various tools used in the industry. . . .

The *Eagle* also recounted one of the grandfather's adventures as a boy:

One of his first adventures in the drilling line was that of drilling a well on his father's property. . . . With the aid of one hundred boys he trailed a large hickory pole from the Frazier farm west of town, and with pole and cable finished the well. . . . His ingenuity and persistence as a boy drilling the water well at home was characteristic of his life in manufacturing well drilling equipment. . . .

Upon completion of his first job, several wells were drilled by him . . . the tools used on this and other wells were made by him in the old Lefever blacksmith shop. . . .

Louise's mother, Loretta Strance Spang, also had an oil background. Strance family members had participated in the early-day drilling of some of the state's most noted oilfields.

So the young woman Henry Keplinger obviously had fallen for was no stranger to oil and engineering, the guiding lights of his professional life.

At Christmas, Louise received an engagement ring from her boyfriend, who knew nothing of Henry Keplinger. And she didn't tell Keplinger about the ring. And Keplinger kept on courting her by long distance telephone. So her romantic life continued to grow more complicated with every passing day.

Then her father and mother decided to attend an American Petroleum Institute meeting in Wichita, Kansas, with a stopover at Tulsa. Louise went with them—and Keplinger was overjoyed at seeing her. The couple had lunch at the Alvin Hotel, and Keplinger asked her to go up on the balcony with him. Once there, he told her that he loved her. "When we get to Wichita, I'm going to ask you something," he said.

Her parents had agreed she could ride to Wichita with Keplinger. On the drive, Louise had no trouble in guessing what Keplinger would ask her, and she spent much of her time trying to think of how she would answer him.

At Wichita's Winfield Park, Keplinger photographed her behind a line of wire trash baskets. "There!" he said triumphantly, "I have you trapped!"

They enjoyed the dancing and merriment at the API meeting, but there came the evening when Keplinger drove her out on the University of Wichita campus. There, parked under a huge pine tree, he said the words she would always remember.

"Louise," he said with grave tenderness, "I have a lot of things to accomplish in my life, and I'd like to have you by my side." It was the proposal of a talented and ambitious man.

It was the moment of truth for Louise. She slid over on his lap. "I'm engaged to somebody else," she admitted, "and I've got his ring." (How true the statement; she was wearing the ring with the jewel turned under!)

"That's no problem," Keplinger said. "You can give the ring back."

But she didn't say she would. "I need some time, Henry. I think a lot of you, and you know it. But I need some time."

And that's the way things were—on the surface—until she went back home. Beneath the surface, she knew that Henry Keplinger of Tulsa, Oklahoma, was the man for her.

She no longer tried to hide from her sisters that she was being courted by Keplinger, and so he wrote her . . . wrote her love letters a poet could have envied.

He developed a custom of sometimes writing half a letter, then phoning her, then completing the letter after they hung up. In late February he concluded the first half of a letter thusly: "I'm going to finish the letter after we've talked. I have hoped and prayed that you will say 'I will,' and I will get a ring and it will be hard to hold me in Tulsa."

During the phone call, Louise, her resistance drained away,

told him she would marry him. In the second part of his letter to her he wrote:

> You have made me the happiest and richest man in the world, and I will do everything humanly possible to deserve your love and honor, and through the grace of God may our love and companionship grow more beautiful each day. . . .
>
> I know that your part in raising a family will be very wonderful, but there are certain hardships which you will have to bear, and I will try at all times to comfort you in times of trouble, and you will always have my undying love. . . .
>
> You sounded so natural over the phone. Sorry your phone connection was weak, but after I knew you would marry me I could not talk much myself. . . .
>
> I'll write tomorrow. May God give us a very beautiful future together. . . .

Twice Louise sent the engagement ring back to the ex-boyfriend, and twice he sent it back. On her third trip to the post office in the small town, the postmaster said kindly, "You're having a hard time getting rid of that, aren't you, Louise?"

Keplinger had planned to arrive in Butler on Louise's birthday, March 25, but some work delayed him. He arrived on March 29 to spend a week. (The final journey of his life was made on a March 29, compounding the significance of the date to the woman who would spend 45 years by his side.) He spent the week in the Spang family home.

The Spangs took him on a trip to the historic oilfields near Bradford and Oil City. Ferdinand Spang took him on a tour of the huge Spang plant, something they both enjoyed. And the entire family took him to their hearts.

The marriage ceremony took place on June 8, the day after Keplinger's twenty-sixth birthday. The vows were given before an improvised altar in the living room of the Spang home. Standing with Keplinger as his best man was his old friend from El Dorado, John Bunyan Adams, Jr.

Louise's veil, according to the Butler *Eagle*, was "a sweeping creation of tulle which fell in a long court train from a tiny Flemish cap rimmed with pearls. . .

"Louise was lovely in a gown of white satin, fashioned on Grecian lines. With it she wore a short jacket of white lace, which was made with long, tight sleeves, and which swept back in a court train. Her arm bouquet was of gardenias and valley lilies."

Later in the evening, the *Eagle* reported, Mr. and Mrs. Keplinger left on a honeymoon motor trip to Canada. "They plan to go from Canada by boat to Chicago, and drive from there to Tulsa, where they will make their home. . . ."

□

They were in Tulsa less than a year. It was a time of adjustment. The bride was determined to "make a home for Henry," but the simple truth was that she didn't know how to go about it. The warm and comfortable environment from which she had come had not provided her the experience. So Keplinger set about "teaching" her. For all of his confidence, he knew little more than Louise. On grocery shopping trips, for example, he would invariably buy the largest package of any product until the engineer in him realized the folly of such purchasing for two people. As for Louise, while she was to develop into an excellent homemaker and a delightful hostess, some of her early assumptions would cling to her for decades. The best example: She served Keplinger soft-boiled eggs at one of their first breakfasts together. Keplinger, a man in love, appeared to relish them. And thereafter, as regularly as the sun rose, he had soft-boiled eggs for breakfast when he was at home. But in 1976 he smiled across the table at Louise and said, "After forty years, I think I ought to tell you—I don't like soft-boiled eggs. . . ."

Shell transferred Keplinger to McPherson, Kansas, in May 1937, and the couple moved into an oil camp nine miles outside the town. Louise was pregnant, and the other wives in camp welcomed the expectant mother to their midst.

The transfer had meant a promotion for Keplinger and greater responsibilities. Nevertheless, he determined to become active in the life of the city. He joined the Rotary Club. He took a Dale Carnegie course in self-improvement. He

helped organize a Toastmasters Club. The citizens of Mc-
Pherson had a pretty low opinion of oilmen in general, and
particularly those who lived in the camps in the McPherson
area. Keplinger was to change that attitude. His friendly manner, his willingness to participate in civic projects, soon made
him a popular figure. So much so, indeed, that he was invited
to speak before various civic clubs in McPherson and in other
communities as well.

In these speeches he was quick to point out that his mother
was from Canton in McPherson County, and that his beloved
grandmother settled in the Canton township area with her parents, Mr. and Mrs. J. B. Griffith, in 1879. So he was one of
them, he was saying.

He had learned an important thing in the Dale Carnegie
course—if people ask you to talk about something they know
nothing about, also tell them something about familiar things
they may not have learned or may have forgotten. So it was
that when he was asked to talk about oil, he also talked about
other industries also . . . and he always tossed in some history
and philosophy.

He knew when the first log cabin was erected in the
county—1866, by C. F. Norstrom, on a claim near the city of
Lindsberg. And when the first baby was born—on January 13,
1869, the son of the cabin builder, christened John K. Norstrom. "He is still living at this time," Keplinger would say,
"and is a resident of California. . . ." Which was news to many.

He had made it his business to learn about the crops and
their beginnings; the coming of the railroads and their growth;
the poultry business and the thousands of eggs and chickens
shipped to the eastern markets annually; how much milk, but-
ter, and cheese the county produced each year. He had learned
the name of every flour mill and how many barrels of flour
each milled annually . . . and he could outline in meticulous
detail the steps which transformed a bushel of wheat into hot
biscuits.

In the same manner he told them about McPherson County
oil, from its origin through the refining process. He knew the
oilfields, their geology, and their potential. He knew the refin-
eries and their capacities. "The greatest natural resources in
present day values," he would say, "are the oil and gas depos-

its. Up to January 1, 1937, 50,000,000 barrels of oil have been produced in the county. During 1936, the production of oil totaled 4,383,519 barrels of oil from seven pools. Six hundred seventy-five wells produced at the end of 1936, and there have been almost a thousand wells drilled in the county. . . ."

And he had learned how to wind up such a speech. "There are many social and cultural advantages, such as good churches, lodges, Y.M.C.A., theaters, swimming pools, good schools and colleges. Recently McPherson College celebrated its fiftieth anniversary, and during its life has been a good benefactor.

"I think the folks of McPherson County are mighty fortunate, and as to the people, I believe they are the ones Bacon had in mind when he said, 'The less people speak of their greatness, the more we think of it.'"

The speech was magnificent, and he had his notes collated because he thought they might be of interest to his children and the grandchildren he was certain would be his.

The speeches he made were so full of interesting facts, so simply delivered by an obviously intelligent, well-educated man, that the citizens of McPherson County could well infer that all oilmen were more than they appeared to be at first blush.

☐

Louise went to Wichita, Kansas, for the birth of their first child. After the new family member made his appearance, the couple used oilfield jargon to let it be known on a simple card. "Amazing New Discovery, McPherson Area; Producers Highly Elated and Happy to Announce Initial Discovery . . . Henry Ferdinand Keplinger; Completed August 25; Potential 6 lbs., 8¼ ozs.; Mr. and Mrs. C. H. Keplinger."

On the card sent to his parents in Tulsa, Keplinger wrote: "I was the toolpusher on this well. . . ."

☐

While still in Tulsa, before the move to McPherson, Keplinger had become deeply involved in petroleum and engineering organizations. He was more than a simple "joiner," however. As an old friend said, "Henry liked to see things run right." So he was always available for the tough assignments, always suggesting ways to better an organization—and always ready to lead the way when his suggestions were accepted by the membership.

He was chairman of the program committee, for example, when the Mid-Continent Section of the American Petroleum Institute held its annual meeting in Tulsa on February 25–26, 1937. It should be remembered that he had been in the oil business less than four years and was only 26 years old when he undertook this demanding assignment. It was his task to obtain the knowledgeable speakers to discuss the various phases of oil production. How well he did the job was noted in the Tulsa *Tribune*. "The excellence of the program being carried out at the A.P.I. meeting has brought forth many compliments to C. H. Keplinger, engineer of the Shell Petroleum Corporation, and who was chairman of the program committee of the Mid-Continent staff," the newspaper reported.

Two years later he was named chairman of the organization when the A.P.I. group met in Oklahoma City.

By then he had been transferred from McPherson to Wichita, and was in charge of production in Shell's Kansas district. And, by then, he had been elected chairman of the Mid-Continent section, petroleum division, of the American Institute of Mining and Metallurgical Engineers.

(His resume in later years would list him as a member of the American Society of Mechanical Engineers, Society of Petroleum Engineers of A.I.M.E., Society of Petroleum Evaluation Engineers, American Gas Association, American Petroleum Institute, American Arbitration Association, Independent Petroleum Association of America, Mid-Continent Oil and Gas Association, Tulsa Geological Society, American Association of Petroleum Geologists, Mexican Geological Society, Mexican Petroleum Engineering Society, and L'Association Française des Techniciens du Pétrole. He held high office in many of the organizations, made significant contributions to their progress, and was singularly honored by them.)

☐

It was in Wichita that Keplinger began keeping what he called "work books." They were work diaries, really, used primarily to describe the work done each day of the year. But during the years with Shell he also jotted down chores he had done around the house, names of people with whom he had lunched or played bridge, ideas that gripped him, books he had read, quotations that interested him, and dreams of the future. He made notes on speeches he heard at clubs, commented on recitals he attended, even described food he had eaten.

He traveled all over Kansas, from one Shell installation to another, with occasional trips to neighboring states, drinking in the sights and sounds and reading everything that fell into his hands. And he made notes in the work books, still misspelling simple words and showing no regard at all for punctuation. He would make side trips to shops and plants, and then draw diagrams in the work books of machines or products he found new and exciting.

While driving to Tulsa, he wrote: "Thought of some admixture such as aluminum to asphalt paving to aid in night driving and to prevent slipping in wet weather. . . ."

On another trip he wrote about Pocket Books: "Started in 1939. Distribute through 700 independent magazine wholesalers in cities over 5,000 population. Retailer makes 6¢ a copy, or 24%. Most popular books: *How To Win Friends, Wuthering Heights, Pride and Prejudice, Lost Horizon, Good Earth.*"

At another time: "Thought of a portable derrick design"— and there was an accompanying crude sketch.

A reminder to himself: "Improve ability to express myself in writing, and coordinate ideas into clear expressions. Smooth English and watch articles." On the same day he wrote, "Avoid 'the above subject,' 'this or foregoing,' 'we find,' 'along these lines.'"

Another crude drawing with the note: "Thought of new locking device with round thread tubing to prevent screw-up."

And there was this: "Be careful in acting too quickly in punishment of Keppy. Learn to observe people critically and describe their actions." Keppy, of course, was Henry Ferdinand, who was growing rapidly and making frequent trips with Louise to Butler where Ferdinand Spang was spoiling him as only a doting grandfather can spoil his first grandchild.

He wrote, too, about Keppy's dog, Mickey. "Fixed crate and

sent Mickey to Butler. Had toilet fixed. Keppy had dropped in toy rubber car."

He was always meeting someone who told him something interesting or amusing. He met them on trains, in diners, service stations, and post offices. "Met a man who told about the discovery of Kotex. Original machines built for Gov. during 1917 War (World) for making bandages. Woman nurse came back with idea."

And on January 29, 1941, he noted that he had received his Draft Classification. It was Classification III, which meant he was not subject to immediate draft. On the same day he "visited with Mr. Cy Roberts from McPherson, who dropped in. He had visited industrial shops in the East and reported capacity operations. Working on a new shell that goes up, then explodes in the sky with a rain of 45 bullets."

And he got a haircut.

These personal notes were interspersed among the notes detailing the great amount of work and study he did each day for Shell. Even when he went to Butler on vacation, he visited steel plants in Pittsburgh and oilfields in nearby counties, and in the Spang plant at Butler he "watched a machinist turn a 3½" pin on a 6¼" round back bit."

Ferdinand Spang wanted Keplinger to move to Butler and manage the company. It was a tempting offer, but Louise said no. She didn't think it would be wise to become that closely involved in family matters. And she believed that Keplinger had a bright future with Shell if he wanted it. More, she knew he nursed a dream that could not be realized in the confines of any company not his own.

☐

In April 1941, Shell transferred Keplinger back to Tulsa. Again, the transfer meant a promotion and greater responsibilities. Now he was specializing in overseeing secondary recovery projects, and his work took him to a half-dozen states as far apart as Kentucky and California. He had long been interested in waterflooding and other secondary recovery techniques, and he plunged into his new work with his usual enthusiasm for learning and doing.

It was not long, however, before Shell sent him to Centralia, Illinois, where an engineer of Keplinger's all-around ability was desperately needed to supervise the company's holdings in that area.

Louise by now was carrying their second child. She went to St. Louis when the time came for the delivery. In keeping with the sex of the newcomer, the card announcing her birth was considerably more decorous than the one which introduced her brother to the world. It was dainty, and it was pink. In effect, it said that Karen Louise Keplinger had entered this world on May 1, 1944.

Meanwhile, Keplinger had been ordered back to Tulsa as Division Engineer for Shell's Oklahoma operations. Leaving the family in Centralia so that Louise could regain her strength, he went to his favorite city to arrange for housing. He was proud of the promotion and happy about the transfer, but he disliked being away from his family so soon after Karen's birth.

In a letter he wrote Louise from Tulsa he said, "I was most pleasantly surprised to have a sweet love letter from you when I went down for breakfast. You won't have to wait much longer, and I am more anxious than you. You and Karen are my favorites, and I want to be by your sides. . . . Karen must be a wonderful little lady by now (she was not quite six weeks old). I will be her constant admirer when I get home. It sounds as if she has some beautiful presents for her Dad to try on her. . . ."

The little girl rounded out the family. There would be no more children. But wherever he went, children would be drawn to him, primarily because he seemed to regard them worthy of his attention and respect.

☐

Keplinger had come far in a short time. There was every reason to believe he would continue to advance through the Shell ranks. But in late September 1944, he told his superiors at Shell he was resigning. He said simply that he wanted to "go it on his own" as a consulting engineer. They asked him to stay on for six weeks to see the completion of a project, and he agreed to do so.

On November 15 he moved into a tiny office in the Kennedy

Building. Actually, it was a room in a lawyer's office, but it had a door to the corridor. The space had been occupied by the lawyer's son, who was in military service. For this "hole in the wall," as Keplinger described it, he paid $60 a month.

He bought $30 worth of stamps, spent $6.88 for stationery and had the Herrick Press run him off $50.69 worth of calling cards.

The Tulsa newspapers ran his photograph in their oil pages with the caption "A NEW CONSULTANT." "C. H. Keplinger has resigned as division engineer of the Oklahoma division, Mid-Continent area, Shell Oil Company, to become an active consulting petroleum engineer here. He will specialize in evaluation work, and work with operators of repressuring outfits. . . ."

So he was in business for himself.

He was 34 years old and in the pride of his manhood. He had not made a leap in the dark. He had been preparing for this day since the summer before he entered El Dorado High School when he told his mother he wanted to be an engineer and work in the oil industry.

CHAPTER FIVE

K eplinger got his first client almost as soon as he got his office door open—E. S. Adkins, the Tulsa County "tagman." Adkins was appointed by the state to sell license plates in the county, receiving a commission on each tag and paying his own office expenses.

But Adkins also was a dedicated wildcatter. He was adept at rounding up leases and putting drilling programs together. He had been wildcatting for years without much success until he brought in Benton field in Illinois near Centralia. And even at Benton field—a good one—he drilled 19 consecutive dry holes before the discovery well came in. His character was tested even further when three offsets to the discovery well were dry holes! The discovery well was on the edge of the field. The fourth offset took him back into the field and on the road to riches.

But for Adkins, even the road to riches had chugholes in it. The landowners had leased the acreage to a coal company, and the coal company had leased to Adkins. Litigation ensued. The question: did the coal company have the oil rights as well as the coal rights? After wearying arguments, the Illinois Supreme Court ruled that the coal company did—and Adkins was home free.

He was head over heels in debt when he discovered the Benton field. He had as partners his wife; his son Eugene B. Adkins who was serving in the Navy; Hubert Howard and Hubert Howard, Jr., investors; and J. V. Howell, his geologist. Howell, however, dealt back much of his interest while the 19 dry holes were being drilled.

56

During the long litigation, Adkins' accountant and financial adviser was a long-time friend of Keplinger's, H. O. Reyburn. A lot of oil had accumulated in the pipeline during the trials, and Reyburn had figured out a method whereby Adkins didn't have to pay taxes on the money the oil represented until he actually received the money. And he developed a process whereby Adkins could furnish bond and take the money as he pleased.

Reyburn had been an accountant for Shell, and had become acquainted with Keplinger in McPherson. They had continued their friendship when both were transferred to Tulsa. Reyburn had left Shell before Keplinger and had set up private practice.

Adkins wanted an expert to take a new look at the geological data he had on Benton field. Reyburn recommended Keplinger. Thus the brand new consultant got his first assignment in an old stomping ground.

In Centralia Keplinger met with Adkins and his lawyer, June Van Keuren, a former judge. He studied the geological material supplied him and gave Adkins his opinion in writing. He also conferred with Eddy Cope, Adkins' field superintendent.

His fee: $300.

☐

Shortly after his return from Illinois, Keplinger took on a partner, Joseph M. Wanenmacher, a veteran Shell engineer who had served the company from South America to The Hague. They would be partners for 22 years, each going his separate way in 1966.

Client Number Two was the Arkansas Western Gas Company of Fayetteville, Arkansas. A friend pushed his way into the tiny office to tell Keplinger that Arkansas Western was looking for someone to do some appraisal work. "Give them a buzz," the friend said. Keplinger gave Arkansas Western a buzz, and got the job.

It called for examining some leases north of Oklahoma City that the company was interested in obtaining. The acreage was on the flank of an old oil field. The structure was there, but the sands were shallow and erratic, in Keplinger's judgment. Arkansas Western was looking for more production than the acreage could yield, he decided. He recommended against acquiring the leases.

The fee again was $300, but the job led to a relationship that was continuing after Keplinger's death. Each year Keplinger's firm would make an appraisal of the company's holdings.

The year 1945 sped by. Keplinger spent more time handing out the firm's cards and "visiting" with prospective clients than he did practicing in his profession. If he drove to Oklahoma City to work for a client, he would visit with as many as a dozen prospective ones before he left town for home. It was a practice he would continue all of his working life—and one he would recommend to anyone who joined his staff. But he went further than that. Once back in his office, Keplinger would write brief letters to those he had visited, thanking them for their time spent with him.

"Henry's printing bill in those days must have been enormous," said a long-time friend. "You know, he'd give one of his cards to a hobo, then try to help him find some work, on the theory that the hobo might some day become a client." The friend smiled. "I'm joking, of course, but it's not far from the fact."

In some months of that first year the firm would receive checks from as many as five clients or as few as two. In April there was no income at all. At the end of the year, the firm had grossed $17,588.89 from professional services. Expenses totaled $8,985.68, leaving a net income of $8,603.21. However, the firm had realized a capital gain of $1,300.09 on a lease bought and sold, bringing the total net income for the year to $9,903.30. Both Keplinger and Wanenmacher had been earning about $500.00 per month when they left Shell's employ.

The next year showed no improvement. Gross income was greater—$22,223.35—but so were expenses—$12,339.94—leaving a net income of $9,883.41. The fact was, business was getting worse with the passage of time instead of better.

It was disheartening.

Let H. O. Reyburn, the CPA who had recommended Keplinger for his first job with E. S. Adkins, pick up the story: "My company had an estate tax return to file for E. B. Peters, who had been a long-time oilman. When he died, Peters held a large number of oil properties, and it was necessary to evaluate them. I recommended Henry for the job, and the executors agreed. We took the necessary papers to Henry's office. It took about six weeks for his firm to make the evaluation, and we paid $5,000 for the job.

"Many, many years later, when Henry was quite successful and an internationally recognized consultant, he stopped me one day on South Boston Avenue. He told me that once I had done him a great favor, and I didn't know it. 'The day you came over with the Peters job is when I'm talking about,' he said. 'The day before, Joe and I had agreed that if something good didn't come in the next day, we were going to close the door and shut 'er down. That five thousand dollars kept us going, and after that job, we started getting all the work we could handle.' "

Indeed they did! Now there was no time for abstractions as Keplinger drove back and forth across the country or flew from coast to coast. The work books now were full of business notations and seldom anything else. Dreaming was out, work was in.

"Never turn down a job," became his motto. He reasoned that if he didn't know all that he should about a project, he could learn, or hire someone who did. (During the Suez Crisis of 1956, when the brief war had left the Suez Canal clogged with sunken ships, he strove mightily to gain a contract to open and then operate the Canal for the Egyptian government whose nationalization of the waterway had precipitated the war. He used what influence he had built up in Washington and corresponded at length with Gamal Nasser, the Egyptian premier, to no avail. "It never occurred to Henry that he couldn't muster the people to do the job," said Kenneth Renberg, an associate at the time. "He wouldn't admit the possibility that he might not be able to handle it. That kind of thinking rubbed off on everybody who worked with him.")

Keplinger worked on well spacing and pipeline systems. He estimated reserves and supervised waterflooding of fields. He testified as an expert witness before courts, commissions, and legislative bodies. He wrote papers that were delivered before industry groups and scientific conclaves and later published in oil-related periodicals. He served as both an officer and ramrod in the various societies of which he was a member. And all the while he continued to pass out company cards to prospects and write letters to clients and friends from wherever he would be.

The firm was growing, and so was Keplinger's professional reputation. The staff was growing also, and more office space had been acquired to properly contain it. Everything looked fine except. . .

Keplinger wanted the big one, the one job that would bring the firm national recognition, the one job that would let the firm take its place where he felt it belonged—in the top rank of consulting engineers.

And the opportunity finally arrived. Few men were ever as well prepared to meet the challenge the job presented.

□

The island began as an annoying sandbar in Long Beach harbor. Over the years it grew and grew as spoil from dredging operations was dumped on it. Eventually it became an island some four miles long and a mile wide, large enough to accommodate a huge Southern California Edison steam generating station, an enormous U. S. Navy shipyard with the largest drydock on the Pacific Coast, and a military airfield.

It was called Terminal Island.

The city of Long Beach also had grown. In its early days it was chiefly a residential community and a year-round pleasure and health resort. But the discovery of oil in 1921 at Signal Hill, a neighboring community, launched an era of phenomenal industrial growth. And then in the 1930s came the discovery of another oilfield, Wilmington, which would become the second largest field in the 48 states.

Scores of companies drilled hundreds of wells at Wilmington, and a forest of derricks sprang up on the mainland. Town lots were leased and drilled on in the race for the oil. The city itself formed the Long Beach Oil Development Company (LBOD) and proceeded to drill on city property. And since the city owned the mineral rights to Terminal Island, LBOD set up rigs on the island's perimeter and drilled directional wells out under the water. It was a drilling orgy second only to the unrestrained activity in the great East Texas field where a new well was being spudded every hour of the day.

Then, on August 14, 1941, the Long Beach *Press-Telegram* ran a short story saying that engineers had noticed a slight "settling" of Terminal Island's east end, as much as four inches in some places. A harbor department consultant, Dr. U. S. Grant, said the settling had ceased, and soothingly cited other instances of slight subsidence elsewhere in Southern California, the newspaper said. Three weeks later the subsidence was ascribed to "reduction of hydrostatic pressure in a subsurface stratum which had been laid bare in the naval base dredging operations."

But Dr. Grant was wrong; the movement hadn't ceased. It was continuing, and at an accelerating rate. By 1944 a large area of the island was sinking, with Edison's generating station appearing to be in the center of the widening depression.

Worried about what might happen to its $170,000,000 installation, the Navy sought help from a former admiral, Frederic R. Harris, who headed a large engineering firm based in New York. The Navy also was disturbed by "talk" that construction of the drydock was responsible for the subsidence. Before the construction had begun, a great amount of water had been pumped from below the construction site.

The Navy told Harris: "Learn what is causing the subsidence. Forecast its future course, if possible. Recommend a remedy, if there is one."

To lead the investigation Harris chose an officer of the company, Eugene Harlow, a civil engineer and an expert on soil mechanics. Harlow and his team studied the most likely theories that had been offered to account for the subsidence—theories that had been propounded by geologists, petroleum engineers, soil experts, and even ordinary citizens.

Some had said that heavy traffic on Terminal Island was the villain. Others insisted that the city's water supply, the Silverado Aquifer, which was only 400 feet below the surface, was being rapidly depleted. The general area was faulted, and some blamed activity along those faults for the sinking. And it seemed perfectly reasonable to many that it was only normal for filled land in an unstable delta area to settle down.

Harlow and his team, after months of research and study, discarded all of these theories. They were convinced, they reported, that the subsidence was caused by losses of reservoir pressures due to the withdrawal of hydrocarbons from the Wilmington field, and that it would get much worse if the withdrawal continued at the present rate. They recommended that the Navy build dikes to protect its property against anticipated water encroachment.

The Navy accepted the advice. Work was begun on a diking system that eventually would cost $11.8 million. Edison, fearful that its generating stations might be swallowed, began spending millions to keep the turbines in operation.

Nothing was done to combat the subsidence, and the land continued to sink while the depression widened.

The oil operators were still busy, and two new oil zones were found in the field, bringing the total to five. The new pay zones prompted increased drilling in the field.

Meanwhile, a rumor made the rounds that the Navy was going to shut down the shipyard, which employed more than 6,000 workers, because subsidence had cracked the floor of the massive drydock. The Navy *was* considering closing the shipyard, but as a post-war economy move. The drydock floor *had* cracked, but the drydock remained operable.

The Navy was concerned more about the subsidence than the rumors. Once again Harris was called in. "Every survey has shown that the subsidence has increased," the Navy said. "We want a new study, a more thorough one. We want a new forecast based on this new-found production. Hire a petroleum engineering firm to help you. Perhaps we can find a way to stop this."

Harris turned the assignment over to Harlow, and Harlow set about finding a petroleum engineering firm. He obtained a list of names from the American Petroleum Institute. He phoned and called on persons he respected for suggestions. One name kept popping up more than any other—Henry Keplinger of Tulsa, Oklahoma. Harlow called Keplinger and invited him to New York for an interview. Keplinger was in Harlow's office the next morning.

Harlow was a few years younger than Keplinger, but their backgrounds of hard work were similar. And they resembled each other physically, mentally, and in their attitudes toward life. Each felt easy in the other's company. They struck a deal on the spot, and on May 5, 1948, they flew to California.

☐

During the 1944 study, oil operators had refused to cooperate with Harlow, denying him information he deemed necessary to his investigation. They had reluctantly supplied him with bottom-hole pressure information only after the Navy had threatened to sue them.

Now, in 1948, Keplinger wanted a lot more information than just bottom-hole pressure figures. He wanted to make a complete reservoir analysis, producing zone by producing zone, fault block by fault block. He needed everything from electric logs to production rates. The oil operators turned their heads to his requests.

But Keplinger was persistent—and persuasive. "He was very pleasant to talk to," Harlow would recall many years later, "but it showed through that he was a man who knew what he knew, knew what he wanted, and knew how to get it."

It became evident that the oil operators didn't want to turn over their data for two reasons: they didn't want to give their competitors the slightest advantage, and they were fearful that the Navy would use the data in a damage suit against them.

Keplinger trudged from one oil company headquarters to another, pledging that each company's data would be kept secret, from other companies and from the Navy. "All we want to do is to get to the heart of this problem," he would say. "Certainly you want that, too. The subsidence is damaging you." This was true; well casings were snapping at an alarming rate as if an unseen hand had twisted them.

Finally the most respected operator in the field succumbed to Keplinger's arguments. He gave up his company data and, one by one, the others followed suit. Keplinger had hired a couple of petroleum engineers to help him, and work began in earnest. While his aides studied oil-water ratios, oil-gas ratios—everything Keplinger considered essential—in every pay zone and every fault block, Keplinger flew to Venezuela. He had recalled something from his days with Shell that he thought might be germane to the Wilmington study. Harlow and his aides, meantime, were giving their attention to the soil—the various formations in the field.

In the 1920s, before the discovery of Wilmington field, Shell,

Gulf, and Lago (later Creole) had waged an all-out drilling war in Venezuela's Maracaibo Basin. Shell held vast concessions on the eastern shore of Lake Maracaibo; Gulf held concessions in the Kilometer Strip, originally a fishing zone that encircled the huge lake and extended from shore one kilometer out into the water. Lago controlled the remainder of the lake floor. While Shell's rigs marched up and down the coast, Gulf and Lago matched them derrick for derrick in the shallow waters of the lake.

By the early 1930s the land mass had so subsided that strong dikes had to be constructed to hold back the encroaching lake. With every passing year the dikes grew in height until they eventually cut off a view of the lake in some places.

Shell officials gave Keplinger all the information he asked for. The land near the shore, they said, had been nothing more than a mangrove swamp when the boom began, and draining it to make the area habitable had caused some subsidence. But that sinking, they said, was little indeed compared to that resulting from the unrestrained exploitation of the oilfields. Keplinger examined their records, studied the area, then flew back to Long Beach.

Seven months passed before Harlow was ready to report to the Navy. The report stated without equivocation that the subsidence in the harbor area was caused by the extraction of oil, gas, and water from Wilmington field, and that it could be stopped only by controlled waterflooding on a large scale. Sea water, chemically treated to destroy algae, was recommended as the flooding agent. Sites had been selected where the water could be injected. But it was pointed out that successful waterflooding would require unitization of the great field.

The report also predicted that further subsidence would create a horizontal movement of the earth's surface.

Oil company geologists, petroleum engineers, and consultants were in almost unanimous agreement that waterflooding wouldn't halt the subsidence. They said, in effect, "That guy from the Mid-Continent may know all about Oklahoma and Texas sands, but he sure as hell doesn't know California sands." Waterflooding at Wilmington would merely increase water production in the wells, they contended.

But Keplinger also had said something that had caught the

attention of a few company presidents. Waterflooding would double the field's productive capacity if it were done properly, he had said. It was something to think about.

[ ]

Keplinger had every reason to be pleased with his work at Long Beach, though his recommendations were, at first, ignored. This denied him the recognition such a thorough and profound study should have brought him. He would have to wait to see his judgment vindicated and hear his praises sung. Meanwhile, a strong friendship had developed between Keplinger and Harlow, based on mutual affection and regard. Shortly after the Long Beach report was delivered, Harlow was responsible for Keplinger receiving his first foreign assignment, a dream the Tulsan had cherished from the day he opened his office door.

But Keplinger kept informed about events in Long Beach, and he was happy to note that two years after his report the Long Beach Oil Development Company had begun waterflooding experiments in one fault block in the Ranger zone, an upper zone that was the least expensive to use as a test tube. Later he learned that initial results were favorable, and that experimenting would continue.

But Long Beach was not through with him. The Wilmington field, as drilled, was on an anticline whose peak was almost directly under the Edison generating station. In 1954 seismic tests revealed that the field was not an anticline but a *double* anticline, with a wing out in the water. Its peak was right in front of Long Beach's city hall, and not far offshore. In an industry hungry for new discoveries, the clamor to drill was deafening.

The Navy didn't need to be clubbed in the head to realize the implications. Again the Navy called on Harris, and Harris called on Keplinger.

The situation had worsened unbelievably. The subsidence area covered some 20 square miles, with the center near the Edison generating station, which had sunk more than 27 feet! Long Beach had not escaped damage, either. A waterfront skyscraper had sunk more than 36 inches, the Ford Motor Company plant 15 feet. Store windows shattered, elevators refused to function, water mains broke.

And, as predicted in the Keplinger-Harlow report of 1948, horizontal movement of the surface was evident. Pipelines broke. So did railway lines. Schools, libraries, and other buildings began inching slowly toward the harbor. And in the harbor area more than 250 oilwells were put out of service by twisted casings. Other wells were shut down while their pumping units were removed so that Christmas trees could be elevated 12 to 17 feet!

A story in the Long Beach *Press-Telegram* was headed: "Is Your Fence Moving! Maybe It's Subsidence." It began: "Subsidence is doing some queer things in Long Beach. For instance, take the Los Angeles-Orange County line. The map shows it's in a certain place—but chances are it isn't. Or if you live east of Cherry Ave. and south of Pacific Coast Hwy, you're going places—not so you'd notice it right now; but the land's moving toward Terminal Island. . . ."

Years earlier *Time* Magazine had labeled Long Beach the "sinking city," and now the label seemed appropriate.

But what about this brand new anticline? What is drilling it going to do to us? the Navy asked.

Keplinger and Harlow had little trouble obtaining information this time around. In the six years since the 1948 study, Keplinger had worked in oil provinces around the globe, establishing a modest international reputation that gained him some credence even among the scientists at Long Beach. Too, the small-scale waterflooding experiments undertaken by LBOD had been impressive. And the 1948 Keplinger-Harlow forecast that the earth's surface would move horizontally toward Terminal Island had proved to be devastatingly accurate.

They studied the new anticline and re-examined the old one. Their report was predictable to a large degree, but startling nonetheless: Unless the new anticline was produced with waterflooding to control pressures, subsidence offshore would reach 45 feet at the center, and a considerable portion of Long Beach, particularly along the waterfront, would be inundated. "A shocking forecast," a radio reporter called it.

Then something occurred that left Keplinger and Harlow flabbergasted. Keplinger never missed a Rotary Club meeting if he could help it. Harlow was not a Rotarian, but he attended a meeting with Keplinger in the nearby city of Seal Beach. In the foyer of the meeting hall, surrounded by curious Rotarians, stood an animated model of an oilfield. Oil would be pumped from the producing stratum in the center of the field, and the overlaying strata would begin to subside. Water would be pumped into the stratum from wells at the edges of the field, and the subsidence would cease and the flow of oil would increase. It was a visual demonstration of how waterflooding would stop subsidence and increase production.

Speaker of the day was Charles S. Jones, president of Richfield Oil Company. He had provided the model, and in his talk he made it clear that he favored unitization of the entire field and a controlled waterflooding program of massive proportions.

Never before had an oil company official or ranking company scientist admitted publicly that there was any relation between oil production and the Long Beach subsidence. But Jones was just getting started. He mounted podiums wherever he could find them to preach the new gospel. After one Jones speech, the *Press-Telegram* said in an editorial:

> Mr. Jones' view is that raising foundations, building dikes and filling sunken areas, as is now being done by the city in the subsidence area, does not get to the root of the problem. . . .
> The solution Mr. Jones advocates as the only real answer is to repressurize the field as oil is withdrawn, this to be done by the injection of sea water. Such a program would have two-fold benefits—it would check subsidence and it would increase oil production. . . .
> An effective repressurization program requires that the entire field be treated as a unit, with all operators sharing in the project and in the benefits. . . .
> The next session of the Legislature should not be permitted to pass without full consideration of this vital matter.

As usual in such "vital matters," nothing came quickly or easily. There were lawsuits, legislative battles, claims and counter-claims, and a great deal of pooh-poohing and nay-saying. It was 1964 before the *Press-Telegram* was able to report that "Subsidence has been completely halted in most of the harbor area and all of downtown Long Beach. Correction of the problem has been credited to the massive waterflood program. . . ."

And in 1965 the new anticline was thrown open for drilling, on the condition that it, like the old anticline, be produced under controlled waterflooding.

It amused Keplinger when a *Press-Telegram* columnist wrote: "We've come a long way, haven't we, in proving that one of the most unique and ornery problems any community ever faced can be solved?

"Why, some of the area out there is actually rising under subsidence remedial measures! That's better than the most hopeful predictions. . . ."

It had been 21 years since Harlow had diagnosed the problem. It had been 17 years since Keplinger had offered the solution.

CHAPTER SIX

Keplinger's first overseas assignment came as a result of his 1948 study at Long Beach. Harlow, as noted, had been impressed with Keplinger's blending of engineering and diplomatic skills. He said as much to Admiral Harris, and Harris at the moment had a project that needed immediate attention. Harris could not have known how desperately Keplinger wanted a foreign job. He had wanted the Long Beach job because he had hoped it would enhance his national reputation. But as he had driven over the highways of Oklahoma, Texas, Kansas, Illinois, and other states, Keplinger had sometimes jotted down in his work book his dreams and plans for the international recognition his brilliance promised and his ego demanded.

The Middle East oil provinces were often in his thoughts. So were Colombia (which he forever spelled Columbia), Brazil, Bolivia, Australia, eastern Venezuela. And there were lands that his intellectual curiosity insisted he visit—India, China, Africa.

Admiral Harris sent him to Turkey.

Though it shared some of its borders with three oil-rich neighbors—Russia, Iran, Iraq—Turkey had only one oilfield, and it was small. The discovery well had been drilled on April 20, 1940, at Raman Dagi, near Lake Van in eastern Turkey. A refinery had been built in the same area, at Batman, to process the field's output. But the Turkish government, through its Mineral Research and Exploration Institute, had some ideas for the future and wanted Harris to evaluate them.

The flight to Turkey was a momentous undertaking and

69

Keplinger duly noted the details in his confusing shorthand in his work book. He obtained his first work passport ever ($11) and a Turkish visa ($1) and took off from New York at 6:30 p. m. on March 21, 1949. The plane, flying at 8,500 feet, arrived at Gander, Newfoundland, at 10:45 p. m. (covering the 1,100 miles in four hours and 15 minutes). He left Gander at 1:27 a. m. and flew all day, arriving in Rome at midnight. He had made stops at Shannon, Paris, and Zurich before reaching Rome. He had mailed postcards (25 total) at each stop.

He reached Athens at 5:30 a. m. on March 23 . . . and could not produce a Greek visa for the customs officials. They refused to permit him to board a waiting plane for Istanbul and Ankara. "Get a visa, if you can," they told him. The next plane for Turkey was to leave at 11:25 a. m., and he wanted to be aboard it. He set out to get a visa, but it was late evening before he had one safely in his pocket.

But he had not wasted the day. As was his custom, he had visited. Athens was swarming with Americans, both military and civilian, and Keplinger mingled with at least a dozen. He noted discussions about a Pakistan oil deal and a carbon black "situation" in the Middle East. He covered the page of the day with addresses and telephone numbers he acquired. And before he went to sleep he mailed a letter containing invoice data to his office. He also mailed one to Louise and another to Clinton F. Robinson, an officer with Frederic R. Harris, perhaps explaining the Athens layover.

Nevertheless, he was at the airport and on the plane for Ankara at 7:20 the next morning. Arriving in Ankara, he noted in his work book that the population was about 230,000, and that the area was the home of the long-haired Angora goat. If he was otherwise impressed with Ankara—and Turkey, for that matter—he failed to note it in the work book. (Later on, on other foreign trips, he would occasionally grow almost lyrical about the sights and sounds and people around him. Certainly Turkey with its long and varied history should have fascinated him. It may have been that he was so concerned about doing well on his first foreign job that he concentrated on the work at hand to the exclusion of everything else.)

For a week he worked at the Institute headquarters with Erzen Berent, chief of the Institute, and two of his top men,

Cemil Tasman and Mozlum Oget. The Turks took turns entertaining him in the evenings, and at a function at the country club introduced him to the U. S. ambassador, Robert Wadsworth.

In an office at the Institute, Keplinger pored over geological data, seismic studies, and other information about untested areas, as well as over logs from the Raman Dagi field and operational reports from the refinery at Batman. Also under consideration were plans to build a pipeline from the refinery to the Mediterranean, more than 300 miles to the southwest.

Keplinger flew to the field and studied operations. He inspected a pumping station on the line to the refinery and made some suggestions. While the party was enroute to Batman to inspect the refinery, traveling over roads that seemed at times to disappear, their vehicle got stuck in a canyon. Keplinger must have been reminded of the vacation he took with his brothers to Mexico City. It took several hours to free the vehicle, and it was the next day before Keplinger could inspect the refinery.

And on the following day he was flown over the proposed pipeline route to the Mediterranean coastal city of Iskenderun. He wound up his work with some conferences at the Institute.

Included in the geological and seismic data Keplinger had studied were some reports on the Garzan area. Like Raman Dagi, it was southwest of Lake Van. It was a likely prospect, Keplinger judged. The Turks were pleased that he agreed with their conclusion. Two years later, on June 6, 1951, the discovery well was drilled at Garzan. Keplinger also approved of the pipeline route to Iskenderun, but the line's construction was not undertaken until the 1960s.

The Turkish scientists also had located a couple of structures that appeared to be suitable for the accumulation of hydrocarbons near Adana, to the north and west of Iskenderun but also near the Mediterranean coast. They wanted Keplinger to examine them. A pilot flew him to the area. The structures were so well defined that they were obvious to Keplinger from the air. After his examination, Keplinger flew back to Ankara and told his hosts the prospects were good ones. The Turkish officials immediately laid plans to drill two deep wildcats on the structures.

In a letter to Louise, Keplinger wrote that he felt certain he could get the drilling assignment if he wanted it, but he had decided that it wasn't a proper activity for his organization. (He also wrote that the newspapers had carried a story about the inspection of the oil properties, but had not mentioned his name. And he said he had dined with an English electrical engineer who had been placed on trial for treason by the Russians in 1923; the engineer had defended himself in the Russian language, and had been freed.)

He flew back to New York via Brussels and London and reported to Admiral Harris. Before the day was over, and before he caught a plane for Tulsa, he found time to visit seven oil companies and banks.

Shortly before his death in 1981, Keplinger mentioned to an interviewer that his first foreign assignment had been in Turkey, but he offered no details. He said with a wry smile, "I guess I didn't have much of an impact on the country. Its estimated oil production last year was only about forty-two thousand barrels per day."

☐

Less than two years after the Turkey jaunt, Keplinger flew to Ecuador when that government asked for advice on construction of a pipeline and some gasoline plants, and on enlargement of an existing refinery.

Ecuador had appeared reluctant to join the brotherhood of oil-producing nations. As early as 1700 the Spanish conquistadors recorded that they recovered oil from hand-dug ditches on Santa Elena peninsula to ease their aches and grease their weapons. This practice of hand-digging wells continued into the twentieth century when great oilfields were being discovered by the drill from Mexico to the Far East. Oil would be skimmed off the pits and transported to a small refinery in containers lashed to the backs of burros.

British residents of Ecuador consistently reported existence of the seeps to their government; eventually Anglo-Ecuadorian Oilfields, Ltd., was formed to drill near them. In 1912, the discovery well Ancon #1 found small production at 1,500 feet.

After that, various oil companies tried their luck in Ecuador with little success until 1923, and later in 1948, when some pro-

Henry, the "Cowcatcher Rider."

At Tulsa University, 1931.

Henry, at far left, with the "Tulsa Foot Warmers."

While at Georgetown University, Henry enjoyed canoeing on the Potomac (1931).

At Gottingen, 1932, with his trusty bicycle (top), and drinking a beer before leaving (bottom).

Graduating from Georgetown University, 1933.

Henry and Louise's wedding picture.

Louise was "trapped" by Henry's camera at Winfield Park.

The Honeymoon Express.

Henry, in 1943, when he worked for Shell Oil Company.

Keplinger's parents, Henry and Edna Keplinger, on their Golden Anniversary in 1958.

Eddy LaBorde and Keplinger in kilts at the 1960 Swingeroo.

Bill Graham and Henry and their phony Swingeroo trophy.

Louise, Karen, and Kep at a bullfight in Bogota, Colombia, 1951.

A language record helped Keplinger learn French.

Keplinger lecturing in Mexico.

Keplinger received a plaque from the Society of Petroleum Evaluation Engineers in recognition of his service as president in 1966.

Henry and Louise at Doral in Miami, 1976.

Henry and Louise on the golf course at Marbella, Spain, in June 1976.

Keplinger, Louise, Karen, and Kep in 1980.

William Mildren, Karen, Louise, Velma, and Kep at Southern Hills Country Club in 1981.

Keplinger and Senator David Boren at a 1979 Washington hearing. Inscription reads: "To Henry Keplinger. With appreciation for your friendship and for the great expertise in the energy field which you have contributed to our nation."—David Boren

Keplinger in Russia. Keplinger is second from left in the front row. To his left are Dr. Yusuf G. Mamedov, Jack Peterson, and Peter Richardson.

Keplinger and some Chinese friends in China.

ductive zones were found in the highly folded and faulted seacoast area. By the time Keplinger reached the country in May of 1951, its wells were producing almost 3 million barrels of oil annually. (By 1980 the annual yield was 84 million barrels.)

He arrived at Guayaquil at four o'clock in the morning and was at work at eight. Guayaquil, a large city near the oilfields in the coastal area, was his headquarters. He visited the existing refinery with an eye to enlarging it. He inspected tank farms and pumping stations. He visited the oilfields. But his chief job was determining if a pipeline could be constructed from the oilfields to Quito, high in the Andes mountains.

Quito, capital of the republic, was 9,350 feet above sea level. It was 164 miles in a direct line from coastal Guayaquil, 297 miles by a winding railroad that had been cut through some of the most rugged terrain on the continent. Keplinger spent much of his time riding a train between Guayaquil and Quito. (He made one self-indulgent side trip to Cuenca where a huge market was devoted exclusively to the merchandising of Panama hats. "Doesn't fit" he noted laconically in his work book a few days later. And he took time out to attend a Rotary meeting in Quito.)

He studied aerial photographs as a further aid. He gathered weather data as a construction guide. Then he explained to the Eduadorans what he thought they should do about the refinery and gasoline plants. As for the pipeline, it could and should be built, he said. Then let's do it, said the Ecuadorans.

Back in Tulsa Keplinger made a deal with C-R-C Engineering Company of Houston to jointly design the pipeline system. C-R-C had just been formed by Herbert E. Fisher, who had been chief engineer of the Stanolind Oil Company pipeline department, and Crutcher-Rolfs-Cummings, Inc., pipeline equipment suppliers. The building of the line eventually was farmed out to Williams Brothers, pipeline constructors. It was a tremendous job. The line was laid roughly parallel to the railroad right-of-way, but still the terrain was rocky, and sometimes it appeared as if the pipe was rising vertically. Keplinger noted in his work book that a pipeliner said, "It's all uphill and hot as hell!"

On his last trip, while returning by plane to Tulsa, Keplinger covered a work-book sheet with notes, and also noted that he met an "English chap" on the plane who tried to interest him in some Venezuelan oil properties. He failed to identify the "English chap" in his work book, a rare omission.

He arrived home on a Sunday. The next day he worked nine hours for five separate clients.

☐

Another "English chap," a charming geologist named Eddy Corps, soon had Keplinger traveling again . . . and this time to one of the lands that had captured his imagination—India. Corps came to Tulsa to attend a gathering of the Nomads, the National Oil Equipment Manufacturers and Delegates Society, whose members had traveled overseas in some oil-related activity. Corps worked for Burmah Oil Company, Ltd., and was chief geologist at the Digboi field in remote Assam province which comprised the mountain-locked valley of the Brahmaputra River.

Even the name of the field—Digboi—was enough to stir a romantic's blood. And the way Corps described the area to Keplinger and others made it appear that simple daily living was a rare adventure. Digboi, he said, was surrounded by a semi-circle of mountains wondrous to behold. On the north were the great snow-covered peaks of the eastern Himalayan Range, rising to 25,000 feet. On the east were the 15,000-foot peaks which marked the boundary between India and China. These peaks, said Corps, were "The Hump" over which the Air Transport Command ferried great quantities of supplies to China during World War II. On the southeast was the Patkai Range, rising to 8,000 feet. The first stage of the famed Ledo Road was built over this range, and wartime convoys rolled along it from Assam to beleaguered China. The village of Ledo was 12 miles southeast of Digboi.

The field, Corps said, had been important to the Allies. It was the easternmost Allied field in operation during the entire Japanese campaign, and was only 75 miles from the Japanese headquarters at Shingbwiyang. Throughout the war, drilling was intensified and production maintained above the peacetime level to provide gasoline and oil for the military units based in the area.

Corps told them that the men who had drilled the discovery

well at Digboi—in 1888!—worked on a derrick thatched with palm leaves and used equipment brought to the drill site on the backs of elephants. Wild elephants roamed the valley then, and still did, and the valley was the last great stronghold of the one-horned rhinoceros.

Oil seeps had been noted in 1828 at Digboi by a party hunting coal. Then, in 1882, crewmen building a railroad examined the seeps and told stories about them in the settlements. The stories eventually reached the ears of Burmah Oil, and the field was opened.

After the meeting, Keplinger invited Corps to breakfast the next morning, and Corps accepted. They were having some problems at Digboi field, Corps said, and he wanted to discuss them with Keplinger.

Over coffee Corps said that Digboi was producing a little less than 2,000,000 barrels annually, and that both "shooting" and acidization in a tight sand in the field had failed to obtain a significant change in production. Paraffination was a big problem. Very low salinity of the formation waters made it difficult to differentiate between oil, gas, and water-bearing sands in electrical logs. Severe losses of circulation occurred in certain sands. Waterflooding was being considered to increase production.

The two men flung ideas back and forth long after their dishes and cups were removed from the table. Finally Corps said, "Why don't you come over and take a look, Henry. I'm sure the company will engage you. They'll want your recommendations in London and Karachi as much as I do."

Keplinger smiled. "I'm waiting to hear from you."

☐

The call came in late August. Keplinger flew to London on September 13, 1951, where he visited with Burmah officials, and made calls at other oil company offices. He found time during the three-day stay to attend a television preview and an oil exposition, and to get his suitcase repaired (54 cents).

He had a one-hour stop in Rome on his way to Cairo. He noted in his work book that he "visited Cairo," but he mentioned no details. In a letter to Louise, Keppy, and Karen, however, he wrote that he arrived in Cairo at night and visited the town's "high lights." He added: "Got up this morning, saw some more—the pyramids, Moslem mosques etc. There are thousands of flies, and I was glad to get out.

"We flew over the desert all the way this afternoon (to Bahrain Island). Saw the Red Sea, Suez Canal etc. Keppy could get out a map and lay a ruler between Cairo and Bahrain Island. There is nothing but desert along the whole route. No pasture land or grain fields. . . ."

At Bahrain Island, the plane developed engine trouble and he had to spend the night. "It is very hot here," he wrote. "120° F. . . ."

He finally arrived at Karachi at 8:30 a.m. on the morning of September 18. At the Burmah Oil office, Keplinger visited Keith Laurie, the manager, and Scott Wilson, the chief geologist. He had read all the reports on Digboi field by noon, and Wilson drove him around the city, with a stop at the Rotary Club. He liked Karachi. He was surprised that it was such a modern city with more than a million inhabitants.

He took pictures of public and government buildings and used up a lot of film at the port, noting in his work book that Karachi was the capital of Pakistan and its chief seaport, located on the Arabian Sea at the delta of the Indus River, and was the country's greatest commercial center. (Years later he would be able to recite these facts so that a listener might well believe that Keplinger spent his weekends in Karachi.)

In a letter to Keppy and Karen he wrote: "They use camels almost exclusively to haul freight. One camel is hitched to a 2-wheel cart. Also they have lots of small donkeys for 'light' loads. Karen would love to have one of the donkeys, just about 2½ feet tall. . . ."

Laurie, the Burmah Oil manager, had Keplinger as a dinner guest in his home. Also there were Pakistan government officials and British diplomats. "I enjoyed the evening and got to eat native foods," he wrote his children. "The two best dishes were rice curry with vegetables and big hunks of fish, and a chocolate dessert. Had an excellent chicken course and a special lentil soup. We started at 8:00 and I had to leave at 11:00 without my coffee, so you see, we all had lots to eat. . . ."

He flew to Calcutta and was shocked by what he saw. Again he was driven around the city by a Burmah official. The opulence so manifest on broad avenues lined with magnificent residences, hotels, shops, and clubs was in raw contrast to teeming slum areas of incredible poverty. He looked at the great temples and the squalid dock sections. "Gold and dross," he noted in his work book. In a letter to Louise, however, he wrote only of the gold. "Went to the best clubs, Bengal and Saturday Clubs, very exclusive and lots of waiters with excellent food, good steaks and fish. They eat very heavy here in the better places with 8–10 course meals. . . ."

He flew to Digboi over vast flooded areas and great tea gardens with hundreds of people picking and elephants working along with them. He was driven to the camp site to be met by Eddy Corps, who promptly took him on a tour of the field, pointing to the surrounding mountains as if he had created them that morning. "Hope we run across a rhino," Corps said with a grin.

Keplinger was the only American in camp. Corps and his geologists and mechanical engineers were British. The rest of the personnel were Indian. The Indians of top rank were mostly young men who had never met an American engineer. All of them, however, had heard of Tulsa; and Keplinger, always in love with the city and always happy to talk about it, reminded them daily that Tulsa was, indeed, the Oil Capital of the World.

There were no communal bunk houses for the top staff. Each member had his own bungalow. Keplinger stayed with Corps. Almost every night of his stay someone gave a party in his honor. He played bridge occasionally, and golfed on Saturday afternoons.

"The Corps had a party for me last night," he wrote Louise, "and invited all of the 'I.P.' people over with their wives. There were two Indian couples, and the women were all dressed up in colored shawls and such, and one of them had a red 'caste' mark on her forehead. . . ."

And he worked. Every geologist and engineer in the field had some knowledge of waterflooding, and they were anxious to try the process under the guidance of an expert. Keplinger studied all the available data of which there was plenty, and decided the process was worth the gamble even in such a strange field as Digboi.

And strange it was. It lay on a long, narrow, faulted anticline. There were three major productive sand groups, each made up of several individual oil-bearing sands, and oil was found in chance permeable streaks and lenses scattered irregularly throughout these sands.

One group, the Bappapung sandstone, had five main oil-bearing sands. Each sand held two or more distinct oil pools, separated by barren or water-bearing areas, for a total of 15 distinct pools. Similar conditions existed in the Nahor group. The Digboi group was different. It consisted of 12 main productive sands, but the pools overlapped so that the entire area in which the sands occurred was productive in some parts of the group, with no nonproductive gaps as were found in the Bappapung and Nahor groups.

Reservoir pressure varied greatly from place to place. Even more varied was the caliber of production. Wells put in production at the same time, from apparently identical thicknesses of the same sand and at the same structural level, sometimes varied in initial production from a few barrels to several hundred barrels.

Even the oil was remarkable; its wax content amounted to almost 16 percent, and its high pour point averaged 95 degrees Fahrenheit.

Keplinger completed his work in three weeks. He recommended waterflooding two pools, each in different productive sands of the Bappapung sandstone group. Since the necessary equipment was not on hand, he undertook to arrange for its shipment from the United States.

Before he left Digboi, he wrote Louise:

> Have had a busy ten days in the Jungle, got lots of pictures on my trips through the field, elephants, tigers and monkeys right in the field. They shot a tiger Saturday. Have had to sleep under mosquito netting at all times, and hope I have not gotten malaria.
>
> I was fortunate in having draftsmen and clerks and stenographers at my disposal. I worked them over the weekend which is something 'new' for the country, and finished my report which I am taking with me for a London conference.
>
> The weather has been warm but not uncomfortable. We

get up at 6:00 and report to work at 7:00. The office closes at 4:00. We usually had tea from 4:30 to 6:00 every evening. I have had tea with both English and Indian families. Then I would work until nine in my room. I stayed with the Corps. We ate dinner at nine and went to bed at 10:00 to 10:30. I usually read before going to sleep, read a book on Churchill, a book on India which was very interesting, and a novel. . . .

He flew home, making stops at Burmah Oil's Karachi and London offices on the way. He arrived in Tulsa on a Sunday, October 7.

And the next day, Monday, he drove to Okmulgee!

☐

While he went about his domestic business, Keplinger kept in touch with the Digboi waterflood by mail. Within a year of his first visit, however, he was on his way back to Digboi to study and make recommendations on a second waterflood. And on this trip he went as an ambassador of goodwill for the upcoming International Petroleum Exposition in Tulsa. And he had some other plans as well.

In London and Paris he visited the offices of oil companies and oil-related businesses and left invitations for attendance at the exposition. He did the same thing in Rome. Then he flew to Algiers for the International Geological Congress where he again distributed invitations.

But because he always looked for work wherever he went, he flew into Iran, landing at Tehran. The country was in turmoil. The year before, the people had overthrown the Shah and had seized the oilfields controlled by Anglo-Iranian Oil Company (British Petroleum).

The trouble had begun over implementation of the so-called "50-50 law," which had originated in Venezuela. The members of the international oil cartel for decades had paid Venezuela, the Middle East, and other oil-producing countries less than ten cents per barrel for their oil; royalties ranged from 7.5 percent to 10 percent, while land tax payments generally were offset by exemption from paying import duties on companies' supplies and equipment. During World War II the Venezuelan government instituted the "50-50 law," which guaranteed the country one half of the net proceeds from oil operations. The companies had fought passage of the law in vain.

The doctrine spread to the Middle East; to Saudi Arabia, where the Aramco consortium was forced to accept the 50-50 terms by the Saudi king; and then to Iraq.

In Iran the oil concessions were controlled by a single company, British Petroleum, 51 percent of which was owned by the British government. The deal British Petroleum offered the Iranian parliament was called unsatisfactory. After strikes and political upheaval, the government nationalized the oil industry in May of 1951.

British Petroleum and the other members of the international oil cartel immediately boycotted Iranian oil, making a shambles of the Iranian economy.

It was into this arena that Keplinger strode in September of 1952. He had two ideas in mind. For one thing, he had learned that the government somehow was selling oil to Italy and Kuwait. He wanted to make a deal to purchase oil and then broker the deal to some oil company outside the international oil cartel. His second plan was to try to sell his services as an appraisal engineer to the state-controlled oil company which was in obvious disarray.

He checked in at the Park Hotel and quickly enlisted the aid of a man named Reboisson. Reboisson saw to it that the press was made aware of Keplinger's presence in Tehran, and a short story to that effect was published. The story made it clear that the "American appraisal engineer" was a person of some importance.

The next day Reboisson paved the way for Keplinger to appear before the Oil Purchasing Board, headed by Kaylem Hassibi. Keplinger stated his case. The board members listened politely and told him to draw up a contract. It seemed a simple matter. He drew up a contract, and the next day presented it to the board.

The contract was not *exactly* what they had in mind, the board members said. It needed to be reworked, but along the same lines. Undaunted, Keplinger went back to the hotel and rewrote the contract. When he delivered it the next day he was told to leave the contract; the board members would study it; he would hear from them.

He saw no reason not to pursue his alternative plan, so Keplinger flew to Abadan and inspected the giant refinery complex, which was operating at a fraction of its capacity. Back in Tehran he refreshed his memory on salient points of Iranian production before the boycott. The helpful Reboisson arranged a meeting with the proper people. Keplinger made his appraisal pitch, and was told to draw up a contract. He did. When he presented it, he was told the contract would be studied and taken under advisement.

Then he waited. He waited until he could wait no longer. He had to get to Digboi on schedule. He left Tehran without hearing a word, favorable or unfavorable. And as had occurred on his trip the year before, the plane developed engine trouble and was forced to land at Bahrain Island. It took him two days to get off the island, but he was greeted warmly nevertheless when he finally arrived at Digboi.

He reviewed operations of the initial waterflood and made a study for another. In his recommendation that the second waterflood be instituted, he wrote: "An unusual provision in the waterflood plans provides for injecting the water at a temperature of 100 degrees for the first month because of the high pour point of the crude oil which is 95 degrees. It is hoped that this precaution will force the oil away from the bore holes of the injection wells in a fluid state without plugging on account of precipitation of paraffin and wax. Higher water injection volume at lower injection pressure should result if the plan is successful. . . ."

His recommendation was accepted, as were others in his long association with Burmah Oil and Digboi field.

Keplinger flew home without stopping at Tehran. It was just as well that he did. Not many months later the American CIA and British secret agents fomented a revolt that ousted the government leaders, and the Shah returned to sit on his Peacock Throne. While the National Iranian Oil Company remained owner of the oilfields and installations, members of the international oil cartel were the sole purchasers of Iranian oil because of their exclusive control of world markets.

CHAPTER SEVEN

Keplinger had hardly kicked the dust of India off his heels before he was on his way again—this time to Colombia, and to a job that would put his company in the forefront of consulting engineers. And it was a job in which he had a special interest, an interest sparked by one of the greatest wildcatters of all time. The opportunity came because his spadework paid off. A New York banker, whom Keplinger had visited on several occasions to hand out cards, told him that the Colombian government was in need of services such as he supplied.

The famed wildcatting team of Mike Benedum and Joe Trees had brought in Colombia's first oil well on a huge concession in the midst of a dense jungle. The concession, covering 2,061 square miles, had been obtained from the government in 1905 by Roberto de Mares, a Frenchman.

De Mares' interest had been aroused by an area east of the Magdalena River and several hundred miles from its mouth where oil gurgled from the ground and stained the foliage of the trees. This area had been named Las Infantas by Spanish explorers in honor of the royal princesses of their native land.

By 1915 De Mares had despaired of ever raising capital to exploit his lease. He went to New York looking for buyers.

Standard Oil Company of New Jersey and several other oil companies displayed interest, but they backed down when they saw what hardships would be in store. Benedum and Trees, however, were not fazed by the inhospitable landscape. In 1916 they formed the Tropical Oil Company and made a deal with De Mares.

World War I was on. The wildcatters couldn't get a drilling

rig shipped to Colombia. Finally they found three old rigs in the Colombian jungle that had been abandoned by an earlier wildcatter. They salvaged enough parts to make one rig and had them hauled piece by piece into the steaming interior.

Drilling began early in 1918. At 80 feet the bit struck oil—a 50-barrel-a-day flow. The well was sunk deeper. On April 20, 1918, the well came in, flowing 5,000 barrels per day from 2,260 feet. More equipment was obtained, and before the year ended two more wells were drilled, both better than the first.

The wildcatters didn't care about owning and producing oil, they just loved to find it. So they sat down to talk with Standard of Jersey leaders. A deal was made whereby a Jersey affiliate, International Petroleum Company (Intercol), took over Tropical for stock valued at $33 million.

Now, in 1953, the life of the concession had expired and total ownership of the acreage had reverted to the Colombian government, the banker said. "They need someone to help them through this transition period," he told Keplinger, "and they're thinking of waterflooding to help increase production in some areas of the field." The man for Keplinger to see was Severiano Cadavid, assistant manager of Empresa Colombiana de Petroleos, the government-owned oil company. And Cadavid just happened to be in Washington, D.C., the banker said.

Keplinger was in the nation's capital the next morning. Before seeing Cadavid, he took time to deliver a letter to the White House in which he explained to President Eisenhower the steps he thought the President should take to solve the Iranian crisis. Like most Americans, he was unaware that the CIA was in the process of fomenting the revolt which would restore the Shah to power. Keplinger's work book notes did not reveal just what steps he thought the President should take, nor did they reveal at a later date whether the President replied. The letter was another bit of evidence that he had a high regard for his own opinions on many matters, and that he didn't hesitate to take them to the highest levels.

He visited with Cadavid in the Mayflower Hotel. Cadavid, who spoke excellent English, quickly made it clear that he had in mind a long-term arrangement, that the job was big and the rewards for Keplinger great . . . if he demonstrated that he could deliver.

Acting on his own dictum of never turning down a job because of its size, Keplinger set about selling himself and his firm to Cadavid. It finally was agreed that Keplinger would be in Bogota, the Colombian capital, in three weeks. He would study all the problems involved, then make his proposal.

As soon as he left Cadavid, Keplinger went to a bookstore and purchased a book which promised that he could quickly learn Spanish by reading its pages. He studied the book on the flight home—and he thought of Mike Benedum, who had discovered the great field Keplinger was now going to see.

He had met the legendary wildcatter in Pittsburgh shortly after he had left Shell and gone for himself. While visiting and handing out cards, he had introduced himself to Benedum. Benedum had invited him to lunch at the Duquesne Club, where he had a private dining room. "Never go wildcatting, Henry," Benedum told him over lunch. "It'll get in your blood, and you'll spend all your time looking for oil. And hell, you never make any money at wildcatting." Keplinger had suppressed a smile; Benedum, he knew, had made millions at wildcatting, and had kept most of them.

But what Keplinger remembered best was the advice Benedum then had given him. "You've set yourself up as a consulting engineer, so be one. Never sell a deal for a client, and never take one. The most important thing you have to offer is your integrity. Never shade it. If you think a reservoir is holding a million barrels of oil, never let a client talk you into saying it's holding a million and a half. I'm sure you're a fine engineer, but your integrity can make you more money than your talent."

Keplinger had lunch on many occasions after that with the old wildcatter, and Benedum would smile and say, "No tarnish, Henry?" And Keplinger would smile and say, "Not yet, sir."

☐

Back home, he learned that his old fraternity pal, Mitchell Tucker, was learning Spanish by listening to Victrola records. Tucker, busy climbing the ladder at the *Oil and Gas Journal*, had been assigned to Mexico City to help an editor set up a new office. Keplinger rushed out to Tucker's home and borrowed the records. Said Tucker ruefully, "Henry learned to speak

Spanish, and I never learned any more than to order breakfast. You know—'dos huevos en tostado.' "

Keplinger did indeed learn Spanish. By the time he reached Bogota he had learned enough from the book and the records to hold a halting conversation. When he left after a month's stay he could speak Spanish with ease, and he was learning to read and write the language.

Cadavid had been right; the job was a big one. Keplinger split his time between Empresa's offices in Bogota and the field headquarters at El Centro. The field was big and rich, with promising unexplored areas. Keplinger realized that his firm would have to train Colombians to drill and produce oil . . . and that he would have to employ 50 or 60 Americans and knowledgeable Europeans for that purpose.

He studied the area geology and reservoir characteristics; in his room at the Hotel Tequendama he drew up plans for gasoline plants; he pondered on the necessity of waterflooding certain sections of the field.

When he was ready to make his proposal to the Empresa board, he wrote his speech in English and, with Cadavid's help, translated it into Spanish. Then he read the Spanish translation before the board.

Years later he told an interviewer, "I don't know if they admired my proposal or my audacity in reading it in Spanish, but we got the job."

The job was to last four years. He would make many trips to Colombia during that period, and many a young engineer underwent his more intensive training at El Centro. Diplomacy was vital to a successful operation because the Colombian workers were jealous of their new status in a nationalized oil company. They were sometimes reluctant to take instructions from a foreigner, and the Keplinger staff members had to ease their way around such situations.

"Henry had faith in us," said Kenneth Renberg, one of the engineers at El Centro. "He gave a man the responsibility for a job and he was on his own. Under such circumstances, if you were shaky when you started a job, you were solid when you got through."

Renberg made the initial waterflood study that Keplinger

had recommended, and the work eventually was taken over by the Forest Oil Company.

In a letter to Louise, Keplinger wrote:

> Went to the Ballet Theatre (in Bogota) as guest of the Forest Oil Company. They had the ballet company from New York with Youskevitch, Kaye, Hightower and Kriza. It was very interesting and well done. The actors almost looked unreal at times. They performed 'Design With Strings,' 'Billy the Kid,' which was very good, 'Nutcracker Suite,' and 'Interplay,' a ballet based on childrens' games.
>
> Much more excitement tonight. The entire city is under war conditions. No one is allowed on the streets and there is shooting and half the army is in front of the hotel. It is a practice program. Lots of planes flying overhead. Will read about it in the morning paper. . . .
>
> Have spent the day studying geology and picking our new locations to drill. Hope to finish this tomorrow. . . .
>
> They have taken me to the Jockey Club and the Gun Club for lunches, very special clubs. . . .

On his many trips to Colombia he played a lot of golf with Cadavid and other Empresa officials. He had learned to love the game back in his caddying days in El Dorado, and he played at every opportunity. At one point in his work book he exulted: "Got an eagle on sixth hole at Centro!"

But Keplinger did more than describe the countries he visited in letters to his family. Because the family was such a close-knit group, he kept them together whenever he could. During one of his long stays in Colombia, for example, the entire family spent a month in Bogota.

On another occasion, when business took him to Guatemala, Louise and Karen accompanied him, and the trio spent a wonderful two weeks traveling around the country and the Yucatan area of Mexico. Keplinger had long been interested in Mayan civilization, and the little group spent days visiting ruins of what had been magnificent edifices.

On the flight down to Guatemala, Karen was appointed Junior Hostess by the flight attendants, and she gravely served the pilots lunch in their quarters. She was 11 then, and the passengers were delighted with the child. One man enchanted her with card tricks, and the crew gave her a deck of cards, 52 views around the world as seen on Pan American flights.

On an excursion to Antigua, Keplinger noted: "Many of the structures were started shortly after 1500 A.D. The city was partially destroyed by earthquake in 1773 and abandoned. Some 60,000 people moved over the mountains and settled in Guatemala City. The extent and magnificence of the ruins that still stand show vividly the grandeur of the city that was already old when Boston was in its birth throes. . . ."

While working in Guatemala City, Keplinger was granted a personal, not business, audience with President Carlos Castillo Armas. He noted: "I was so busy trying to speak Spanish that I did not observe much."

He spent a lot of time with Karen. After one sightseeing jaunt he took her to Tullipas to swim in the underground river. In the crystal water, he wrote, "Karen looked like a mermaid swimming with the fishes."

In the States the family went on fishing trips to Michigan, Minnesota, and Florida.

They were happy being together, whenever it was or whatever they were doing.

☐

Another facet of Keplinger the family man appeared in notes he jotted down for a letter he planned to write Keppy. The lad was then 17, and much preferred to be called Kep, but the family continued to call him Keppy. The notes in the work book said:

1. Your life is what you make it.
2. Learn to concentrate.
3. Don't give up.
4. Learn the meaning of discipline and gain self respect.
5. Stand at the head of your class. [Keplinger had written, "I did," but had struck through the additional words.]
6. Appreciate meaning of integrity.
7. Watch your posture & dress.
8. Make decisions promptly; don't procrastinate.
9. Be thoughtful and considerate of family & friends.
10. Be sure to give first place in your life to the spiritual

side. Attend church every Sunday, no matter where
you are. Have communion with God, and you also
will make many friends. Read Sermon on the Mount
once a month.
11. Social life—you will have girl friends, be considerate
of them. Don't get the big head.
12. Value of imagination.
13. How important is money.
14. Be an optimist.
15. Pay heed to conscience.

Read this letter the last day of each month. I ask this
because the things mentioned will be better understood as
you develop. Love & affection.

It is likely that he never wrote the letter from the notes. If he
did, Kep didn't receive it.

☐

Keplinger had not been neglecting his domestic clients as he
journeyed from one foreign land to another. His staff had increased and so had his office space. His clients included the
largest oil companies in America and the smallest. Cities, counties, and states sought his advice on engineering matters.
Banks, foundations, insurance companies, investment firms,
and institutions relied on his judgment in their financing of oil,
gas, and mineral deals.

Such was his reputation for probity that an oil operator could
get a production loan from almost any bank in the land simply
by displaying a reservoir appraisal bearing Keplinger's signa-
ture. At a social gathering Keplinger was introduced to a
prominent eastern financier. Later the financier was heard to
say, "I was glad to meet Henry Keplinger in the flesh. I always
thought he was a foot high from the conservative way he esti-
mates reserves."

Keplinger's fees had increased as the firm had grown, but
many a note in his work book said "no charge" beside a client's
name. Some of the clients were old friends who were down on
their luck and fighting their way back up. Others were engineers who had problems they couldn't solve but didn't want to
admit it to their superiors. Keplinger advised them and kept
quiet about it. Of one such engineer he noted in his work book:

"Waterflood. He was overcharging the formation and the water was just shooting through. No charge."

The part of his work he seemed to love the dearest was testifying at an adversary hearing. The competitor in him enjoyed the intellectual combat. Not that the position he represented always won, but his services as an advocate were eagerly sought. One such hearing occurred at Bismarck, North Dakota, where the State Industrial Commission listened to testimony on which to base a well-spacing judgment for Tioga field. At the time of the hearing, the field was producing on a 40-acre spacing program. Keplinger's client, Amerada Petroleum Corporation, was in favor of 80-acre spacing. It will be recalled that Keplinger worked for Amerada during his college days.

Lloyd W. Sveen, oil editor of the Fargo *Forum*, wrote:

> Oilmen attending the spacing hearing in session here are high in their praise of the study of Tioga field made by Henry Keplinger, independent petroleum engineer of Tulsa.
>
> Keplinger is no stranger to a North Dakota witness stand, for he was one of the petroleum engineers who estimated reserves of gas in Tioga and Beaver Lodge fields for North Dakota Natural Gas Transmission Company in its application for a natural gas pipeline certificate.
>
> Keplinger has assembled into documentary and exhibit form a condensation of every scrap of data possessed by Amerada Petroleum Corporation, the state geologist's office, and other companies pertinent to the field.
>
> Bearing in mind that the Industrial Commission members, landowners and many spectators are laymen, he has written his conclusions in layman's language as much as possible.
>
> While this has aroused admiration of many laymen at the hearing who call Keplinger's testimony an invaluable short course in oil, the highest praise comes from oilmen for the scientific manner in which the data has been assembled to prove Keplinger's contention that one well can drain 80 acres efficiently.
>
> Just before the hearing recessed, H. F. Beardmore, production manager for Warren Petroleum Corporation at Houston, took the stand to concur with Amerada's recommendations for 80-acre spacing. Beardmore is a veteran

of many a spacing hearing himself, and is a highly-re-
garded engineer. Saying he had studied thoroughly Keplinger's prepared testimony and exhibits, Beardmore said:
'The information is the most impressive and most con-
clusive I have ever heard. In most hearings the best an
engineer can do is to present data about permeability and
porosity, and from that try to determine proper well spacing, but here the figures on interference between wells indicate the wide drainage in Tioga field. It is the greatest
wealth of data I have ever seen for a hearing of this
type. . . .'

At the conclusion of the hearing, after all the witnesses for
40- and 80-acre spacing had testified, the Industrial Commission decided in favor of 80-acre spacing. Without doubt, Keplinger's testimony had been most persuasive.

☐

On the Labor Day weekend of 1954 the City Management of
Wichita, Kansas, found it necessary to use police and fire department sound trucks and all local press, radio, and television
facilities to ask public cooperation in reducing daily water consumption by almost one half. The immediate reason: the pipeline bringing water into the city was out of operation, and
emergency wells were, for all practical purposes, completely
dry.

But Wichita's water troubles went deeper than that. The chief
source of the city's water was the Equus Beds, an ancient underground storage basin that extended northwest to Mc-
Pherson. The city had drilled 35 wells into the reservoir. They
were producing 48 million gallons of water daily which was
pumped through a 48-inch pipeline to a city filter plant. But the
city's peak demand for water was 61 million gallons daily, and
demand was growing steadily. Some of the need was met by 14
emergency wells outside the Equus Beds—when the wells
weren't dewatered.

The city had plans to drill 20 additional wells into the Equus
Beds and to construct a 66-inch pipeline to replace the 48-inch
line, which seemed to break every other weekend. A second
stage of the plans called for extending the new pipeline to the

Kanopolis Reservoir on the Smoky Hill River. And there were proposals to drill 40 additional wells into the Equus Beds under certain conditions.

The reaction to these plans in the Newton, Sedgwick, Moundridge, and McPherson area was prompt and vigorous. The Central Kansas Water Conservation Association, made up of about 3,000 farmers, was organized. This group then became part of the Equus Beds Study Group, comprising an additional membership of some 30 towns, villages and municipalities throughout the area north and west of Wichita.

The group contended that the area they represented also depended on the Equus Beds for survival, and maintained that the reservoir already was being depleted too rapidly by the city. The group hired Keplinger to study the problem and present its case at a hearing in Wichita.

Keplinger and his associates made their usual thorough study, and Keplinger appeared before the Mayor's Advisory Committee in Wichita as the group's advocate. He was quick to point out that he was a native Kansan, that he had spent a goodly number of years in the Wichita area, and that he was familiar with Wichita's water problems as a consumer who had drunk his share of bottled water in the past when Wichita's water had hardly been fit to drink. That, he explained, was before the Equus beds had been drilled.

In studying the problem, he said, he had utilized four considerations: that run-off water should be used as much as possible for a water supply source; that annual consumption of water should not exceed the annual replacement from a water supply source; that the rights to appropriate water must not exceed the "safe yield" of water from an underground water reservoir; and that the most economical overall financial operation should be considered.

Pumpage from the Equus Beds, he said, was estimated to be 108 billion gallons on January 1, 1955. (The hearing was in February 1955.) "The withdrawal of this large amount of water has resulted in a substantial lowering of the water levels in the vicinity of this water well field," he said. "On September 1, 1954, when the withdrawals amounted to about 105 billion gallons, the static fluid level had been lowered as much as 32 feet in some of the wells. The area where the water levels have been lowered by Wichita's water wells includes approximately 42,000 acres.

"The large withdrawals of water and the resulting lowering in water levels have had an adverse effect upon other users of water in the area. The most conspicuous result is the large number of dead trees in the well field area. If you care to do so, you may observe this effect in the spring and summer when the trees normally have leaves. The various adverse effects may be enumerated as follows.

"Practically all cottonwood trees have died and many Osage hedge trees are dead. . . .

"As a result of lowering the water level, it had been necessary to deepen many of the wells from which local residents obtain their water supply.

"The quality of water now obtained is much poorer than before the water surface was lowered. The water now contains sufficient iron to cause a pronounced coloration. It is undesirable for domestic use, and especially for laundering, because washed articles are stained. Wichita eliminates this objectionable feature by treatment of the water, but does not provide this service for the local users.

"The area is now not suitable for the growing of certain crops, such as alfalfa, which are dependent upon root contact with the underground water, and as a result the agricultural prosperity of the area has suffered.

"Kisiwi Creek, in the Wichita Well Field, formerly flowed throughout most of the year and furnished water for cattle and other livestock. It is now dry except during heavy rains and for a few hours thereafter.

"The well field is in an area where irrigation is feasible. In fact, several irrigation wells have recently been completed at the eastern edge of the field. The results of irrigation during the past season are reported to have been very successful. Lowering of the pumping levels such as has occurred in the Wichita Well Field results in higher operating costs and reduced yields of wells used for irrigation as compared to areas not subjected to pronounced withdrawals for purposes other than irrigation. This economic factor probably eliminates the irrigation possibilities in the areas of pronounced drawdown of water levels. . . .

"It is our opinion that the withdrawals from this water field at the present rates, or increased rates, will result in continued lowering of water levels in the future. . . ."

He added that it was his opinion that in the past the use and consumption of water from the Wichita Well Field had greatly exceeded the "safe yield" from the reservoir, and would continue to exceed the "safe yield."

If the plan to drill 60 additional wells was followed, he said, Wichita's future source of water would be entirely from the Equus Beds. And this proposal was viewed with alarm by other users in the area.

Wichita could forget obtaining water from the Kanopolis Reservoir, he said. "It is our understanding that Wichita has no rights in the water impounded in this reservoir. When Congress appropriated the funds for its construction, the future usage of the water was dedicated to other purposes.

What Wichita should do, he said, was to look to another source of run-off water supply, "which I know is available through the construction of dams on the Ninnescah River." One dam would be on the north fork of the river, and it should be enough, he said. The water was of good quality. The dam would create what would become the Cheney Reservoir. Total cost of the project as estimated by the Bureau of Reclamation, would be $22,500,000, which included a transmission line to Wichita. That, he said, was considerably less than the cost of alternative proposals.

And he had a little warning for his listeners. "Another factor which must be considered as part of the operating cost, or in the total cost, is claims of damage for use of the water. The construction of the Cheney reservoir and its use as a water supply would not involve litigation and damage claims. On the other hand, the expansion of the Well Field in the Equus Beds could result in expensive litigation and substantial damages resulting from lowering of the water levels, and other adverse effects of large withdrawals from the Equus Beds. . . ."

He wound up with a flourish. "The Ninnescah would assure a permanent, reliable, and uncontested source of good water, not only for the present, but for the future growth of Wichita . . . thereby preserving the remainder of the Equus Beds for Wichita trade territories, to whom they rightfully belong."

It is difficult to assess the impact of Keplinger's arguments on the ultimate decision. One of Wichita's leading citizens, A. E. Howse, was chairman of the Mayor's Advisory Committee on water matters. Howse was a mover and shaker in the very best sense of the term. He obviously was determined to obtain an adequate water supply for the city he loved, but not at the expense of others. It seems likely that he took the best features of all proposals, then forged ahead.

Only 20—not 40 or 60—additional wells were drilled in the Equus beds. The worn-out 48-inch pipeline was replaced by a 66-inch line. Time was not wasted in trying to obtain water from the Kanopolis Reservoir which, as Keplinger had pointed out, had been constructed for other purposes. But Cheney Reservoir was constructed on the north fork of the Ninnescah River, as Keplinger had recommended. And a Ground Water Management District was created to maintain the overall integrity of the resultant water supply system.

With prudent management, the Equus Beds were supplying two-thirds of the city's water in 1982 with the remainder coming from the Cheney Reservoir.

Late in his life, Keplinger told an interviewer that A.E. Howse was "an inspiring man."

□

Though Keplinger was considered a modest man by most people, he once declared unequivocally that he was a "world expert" on helium. The assertion appears to have been sound, for he worked as a consultant and appeared as an expert witness in one of the most complex cases ever to go to trial in the Mid-Continent—and helium was the *pièce de résistance.*

Helium is a colorless, odorless, and tasteless gas which, in the present state of scientific knowledge, will not react chemically or physically with any other element, except under laboratory conditions. The accepted theory is that helium is formed by radioactive disintegration within the earth and migrates to the same reservoir traps as do other gases. Throughout time it has commingled and become diffused with natural gas hydrocarbons which were formed from primordial organic matter.

Helium is noncombustible. It is the second lightest element found on earth next to hydrogen, which is highly combustible.

It does not liquefy at standard atmospheric pressure until it reaches −452.1 degrees Fahrenheit, almost absolute zero and the lowest temperature of any gas.

Helium was first discovered in the spectrum of sunlight in 1868. In 1905 it was discovered in gas from a well near Dexter, Kansas. But it remained a laboratory curiosity until 1918 when the United States began commercial extraction of helium from gas produced in Petrolia field in Texas for use in military balloons and blimps in World War I. With the Helium Act of 1925, the government placed the production and marketing of helium under the Bureau of Mines.

Until World War II, the uses of helium outside of the laboratory were in lighter-than-air craft and, when mixed with oxygen, in medicine and deep-sea diving. Much of it was produced at government plants. But amendments to the Helium Act of 1925, enacted in 1937, directed the Secretary of the Interior to buy the only two private plants in operation at that time. This was done.

During World War II the government built four more plants, but shut down all but one when the war was over. Then, in the 1950s, demand for helium soared. It was needed for atomic weapons, military and civilian rocketry, nuclear energy plants, space exploration, cryogenics, and a dozen other uses. Because of its unique properties, visionaries saw it associated with a future energy source approximating the miracle of perpetual motion.

The government was aware that measurable amounts of helium could be found in most of the natural gas fields in the country. But it also was aware that about 99 percent of the helium resources were in the natural gas reserves of the vast Hugoton field, which covers approximately 33,000 square miles and more than 21 million acres in three states—Kansas, Oklahoma, and Texas. In this great field the helium content averaged between .4 percent and .5 percent in volume. The economically recoverable helium was estimated at 119 billion cubic feet from estimated resources of 36.4 trillion cubic feet of gas.

This helium was being wasted. No matter how many components were removed from the natural gas—butane, propane, gasoline, and the like—helium was not affected by the processes. It remained in the pipeline with the carrying gas. When the carrying gas was burned in hundreds of thousands of stoves and industrial plants across America, the helium, being noncombustible, simply floated into the atmosphere where it could not be captured for use.

In an effort to conserve as much helium as possible, amendments to the Helium Act were added in 1960. They authorized the Secretary of the Interior to enter long-term contracts for the acquisition, processing, and transportation of helium. Also, the Secretary was given the power to acquire helium by eminent domain proceedings if he were unable to obtain it on reasonable terms and at a fair market price.

As in most other oil and gas fields in the country, the landowners in the Hugoton field area had leased their acreage to producers who subsequently drilled for the natural gas, produced it, then marketed it to large gas pipeline companies which transported it interstate for resale.

In the early 1960s four major pipeline companies contracted with the Bureau of Mines to sell helium to the government. They each formed subsidiaries for this purpose and constructed extracting plants to supply the government with a noncombustible mixture of helium and nitrogen. The government removed enough of the helium to take care of current needs and stored the remainder of the mixture in a giant West Texas reservoir.

Producers sold natural gas from the Hugoton field to the pipeline companies for 11 to 15 cents per 1,000 cubic feet (Mcf), and the landowners received one-eighth of that sales price.

The pipeline companies sold the helium-nitrogen mixture to the government for $11 to $12 per Mcf, the value placed on the helium alone, and peddled the remaining fuel gas at its regular price to their regular customers, who found that the gas burned better without the helium in it.

The government, in turn, extracted the helium from the mixture and sold it to federal agencies and selected private companies for $35 per Mcf.

News of this kind of price structure spread across Hugoton field and among the 30,000 landowners who were receiving one-eighth of 11 to 15 cents per Mcf for the helium-containing gas under their acreage. (The government and the pipeline companies, labeled the Helex companies, would later contend that every step in their dealings was well publicized.)

During negotiations between the government and the Helex companies, some thought had been given to the possibility that others would claim title to the helium. It was agreed that the government would accept all risk of title except for the first $3 per Mcf, which would fall to the Helex companies.

The landowners sued the Helex companies in state district court. The Helex companies quickly countered by going to federal court for a judgment. They sued both the producers and the landowners, and the government entered the case on its own behalf. A fund was created which represented the money the Helex group expected to receive for the helium over the lifetime of the various contracts with the government. The Court was asked, in effect, to decide to whom payment for the helium belonged and how much would have to be paid.

The case would become so involved and have so many ramifications that it would remain unresolved even after Keplinger's death in 1981. But the basic arguments at its inception were simple. The landowners claimed that helium did not pass to the producers under the existing gas leases. The producers claimed that the helium did indeed pass to them under the existing leases, but that it did not pass from them to the Helex companies under existing sales contracts. The Helex companies claimed that the helium did in fact pass to the producers under the existing leases, and that it passed on to them under the existing sales contracts. The government concurred with the Helex group position.

Enter Henry Keplinger. He had been fascinated by helium since his days at George Washington University and his advanced studies in physics. It had intrigued him that helium, when liquefied, assumes a physical state unlike any other known to science, a superfluid capable of climbing up the outside of a glass in defiance of normal forces of gravity. In this state it is sometimes described as a fourth state of matter, neither gaseous nor solid nor liquid in the normally understood sense of the words.

Further, Keplinger had done work in 1952 for the Bureau of Mines at Dexter field in Kansas where helium was first discovered coming out of a well in 1905.

Perhaps more important, he was remembered by the landowners and their lead attorney, Dale Stucky of Fleeson, Gooing, Coulson and Kitch of Wichita, for his testimony—pioneer testimony—in an earlier set of cases dealing with the relationship between the price of gas and its use and conservation. Even that far back he was warning the country that it would some day face a shortage of hydrocarbons, a warning he would repeat with monotonous regularity as the years flew by.

As a consultant and expert witness for the landowners, Keplinger's primary task was to convince Judge Wesley E. Brown that the judge could, indeed, try a case that involved millions of acres, thousands of landowners, hundreds of producers, the Helex companies, and the U.S. government. Judge Brown's concern: if he decided that the landowners or producers were entitled to recovery, how could the value of the helium be accounted back to the wellhead? Who in the world could figure out who was to get how much?

Keplinger explained that the problem was handled routinely every day in the natural gas business in connection with the recovery of liquid petroleum products such as gasoline, butane, and propane in processing plants which receive natural gas from a large number of wells after gathering. Through engineering and accounting procedures well accepted in the industry, the plants account back to the wellhead the amount of liquid products recovered.

Computers were just then being placed in general use, and Keplinger, who had studied them, explained how these devices could handle the accounting job even more efficiently.

Convinced, Judge Brown ruled that he would try the case on a consolidated basis.

That was the only thing the landowners won. After hours upon hours of wearying testimony and cross-examination, Judge Brown held "that in view of circumstances existing at time of oil and gas lease executions, definition and use of terms within the industry, and by intention of parties as disclosed by their actions, grant of 'gas' extended to entire gas stream which emerged at wellhead absent express reservation of any constituent elements, and helium passed thereunder unless expressly reserved; and that gas purchase contracts were construable as contracts for the purchase of fuel, not energy or Btu's in view of plain language providing for purchase of gas, and such contracts conveyed lessee-producers' title to gas and helium therein to pipeline companies."

In other words, the helium belonged to the Helex companies.

The landowners and producers appealed this decision. It was March 2, 1971, before the appeals court delivered its opinion. It ruled that the producers should share in the profits from the helium sales and, in turn, should pay the landowners a proper royalty thereon. It ordered Judge Brown to figure out who should get how much.

Before Judge Brown could make a determination, another helium case was decided in an Oklahoma federal court by Judge Luther Bohanon. The suit was brought by Ashland Oil, Inc., against Phillips Petroleum Company—Ashland as a producer, Phillips as a Helex company. In arriving at a decision, Judge Bohanon utilized a formula that Keplinger had propounded during the Kansas hearing before Judge Brown— "proceeds less expenses."

And his decision was directly the opposite of Judge Brown's.

The helium picture had changed considerably over the years, and helium was selling on the competitive wholesale market for $20 per Mcf. Judge Bohanon deducted the expenses for extracting and handling the helium to arrive at a wellhead value of $15 per Mcf. And he ruled that Ashland was entitled to recover $15 per Mcf for every whiff of helium Phillips had extracted from Ashland's gas, plus interest and attorney's fees. And though the landowners were not a part to the suit, Judge Bohanon ruled they were entitled to one-half of the proceeds.

Under Phillips' agreement with the government, Phillips would have to pay $3 and the government $12 per Mcf to Ashland. Phillips and the government appealed. The case drifted up to the Tenth Circuit Court of Appeals, the same appellant court that had reversed Judge Brown in the Kansas case.

Then Judge Brown rendered his second decision. The "proceeds less expenses" formula propounded by Keplinger at the first trial and utilized by Judge Bohanon in Oklahoma was again rejected. Instead, Judge Brown ruled that the producers should get 60 to 70 cents per Mcf for the helium, and the landowners should get 50 percent of that as royalty.

Again the landowners and producers appealed, and the higher court heard arguments in the Kansas and Oklahoma cases at the same time. After many months, the appeals court decided that the "proceeds less expenses" formula should be used in the Ashland-Phillips case, but that the landowners there were only entitled to the royalty fraction set out in their original gas leases—usually one-eighth. But, as was pointed out earlier, the landowners had not been a party to the original Ashland-Phillips suit and were not involved in the appeal.

Judge Bohanon tried the Ashland-Phillips case again. In his decision he once again used the "proceeds less expenses" formula, but with a twist. Instead of starting with the competitive wholesale price for helium—$20 per Mcf—as he had done in the original suit, he used as a starting point the price the government agreed to pay Phillips for the helium-nitrogen mixture under their contract—$12. And he decided that the proper value for all the helium involved over all the years was $3 per Mcf. Three dollars, it will be recalled, was the figure beyond which the government would not have to pay third-party claims.

The government and Phillips appealed. So did Ashland, on the grounds that the appeals court, in approving the "proceeds less expenses" formula, had not disturbed and had approved as a starting point the $20 competitive wholesale price of helium.

But the appeals court affirmed Judge Bohanon's decision, and the United States Supreme Court refused to review the case—apparently making that decision final.

But the Kansas landowners and producers were not through. On reargument of their case in 1981, they took the position that as in the Oklahoma case—*Ashland vs. Phillips*—the "proceeds less expenses" formula had to be used in the Kansas case as well, with the starting point the $20 wholesale competitive price of helium during the time the helium was taken, and with a 50-50 apportionment of the remaining value to be made between the landowners and the producers, as Judge Brown had ruled at his second hearing.

There had been no decision in the case at the time of this writing, but there is no doubt that the Kansas landowners wish that Keplinger was still in the arena with them.

Said Attorney Stucky: "It is frequently most difficult for landowners in litigation such as this to obtain convincing and authoritative advice and testimony from within the oil and gas industry—and that is where most of the expertise is found. We have found that many or most of the best qualified oil and gas consultants have been very reluctant to testify as to the facts or give their opinion in court if requested to do so by landowner parties. Many such consultants seem to be afraid of alienating their principal source of revenue—members of the oil and gas industry.

"But not Henry Keplinger. He was widely respected as a consultant and engineer within the oil and gas industry both in this country and abroad. He was employed by majors and independents alike, and testified on their behalf on many occasions.

"But this did not prevent him from making his expertise available to all, including regulatory authorities and domestic and foreign governments.

"He called the shots as he saw them, and felt free to consult in proper cases with landowners even when they were engaged in litigation with members of the industry, avoiding, of course, direct professional conflicts of interest and always protecting proprietary information.

"His expertise was invaluable to us, and his personal stature in the industry was of such magnitude that he had no fear of alienating his other clientele.

"Henry Keplinger stuck with us all the way until his death. He is sorely missed by the many small landowners and farmers who had come to rely upon his wisdom and advice throughout the years."

CHAPTER EIGHT

A s the years passed, Keplinger testified in scores of hearings before commissions, boards, committees, and juries in a score of states and in Washington, D.C. In every instance he was accorded the respect his background and demeanor warranted. But that was before he became involved in the "Doris Day" case. On a memorable day in a Los Angeles courtroom throbbing with heat because the air-conditioning system had failed, Keplinger was subjected to a grueling cross-examination that sometimes rattled his composure.

Further, the judge implied that Keplinger talked too much, and also made him empty his briefcase so its contents could be studied by his cross-examiner. Keplinger had taken the briefcase to the witness stand because it held material pertinent to his testimony. "When a woman comes to the stand with her purse, and she uses it in connection with her testimony, counsel is entitled to look through it," Judge Lester Olson explained to the bewildered Keplinger.

Over the years Keplinger had established a set of rules to govern his testimony at hearings, and he had taught them to his associates. There were four:

1. The best witness must not appear too zealous for the position of his client. Good witnesses are only concerned with presenting the facts that bear upon the question before the court; they should not appear to be making an effort to persuade the court to accept their views on the question but rather to present the facts in such a manner that the court cannot fail to see their

significance. Put in another way, the *facts* should persuade the court, not the witness.

2. Avoid drawing conclusions from each exhibit presented unless the court requests you to do so. Reserve the conclusions for a final statement, which should only be given in response to a question propounded by your own attorney. (Of course, you should have prepared this question for him in the pre-trial study.)

3. Present or admit evidence against your side of the case before your opponents do so, if you can possibly do it. This enables you to minimize such evidence rather than attempt to combat it.

4. Under cross-examination do not appear reluctant to admit an error of calculation or judgment. Do not let your voice weaken at such a time, but be utterly frank and open about it.

These were good rules, and they had served him as guidelines for three decades. But in that sweltering Los Angeles courtroom he appeared at times to have forgotten them all.

Keplinger's involvement in the case began on December 19, 1968, when he received a letter from Nick Van Wingen, a California consulting petroleum engineer, asking that Keplinger appraise a number of leases in four Texas counties to determine their fair market value. Doris Day Melcher and the estate of her late husband, Martin Melcher, held interests in the leases.

On January 15, 1969, Keplinger reported his findings. "We have now completed this study and find that none of these leases has any value at this time." With that cryptic beginning, he helped the Court in its judgment of Jerome Rosenthal, who had been attorney and business manager for Doris Day and Martin Melcher for more than 20 years. The report continued:

> McMullen County, Texas—Horton Lease. The Horton lease, comprising 1,085.96 acres out of the Horton Ranch, originally had one producing oil well. We have been advised that this well was plugged and abandoned several years ago, and consequently the lease has no value.
>
> Calhoun County, Texas—Various Leases. The eight leases, which are a part of the Carancahua Beach Subdivision and are in the Appling field, have been plugged and abandoned. These leases have no value.

Smith County, Texas—Lydia Wathaw Lease. The Lydia Wathaw lease was originally drilled by Tri-Con and adjoins the Good Omen (Pettit) Unit operated by American Petrofina Company of Texas. However, the well was never brought into the unit since it did not have sufficient production to qualify for inclusion. The well has produced a total of only 8,095 barrels of oil and has not produced since 1964. The lease has no value.

Marion County, Texas—Marion Lease. The properties in Marion County, Texas, were formerly operated by Wilson Oil and Gas Company. The oil and gas leases in which the Melchers had an interest expired in 1962 for lack of continuous production.

A month later Keplinger had other information to report. The properties in Calhoun County had not been plugged and abandoned as he had previously written, but they were worthless nonetheless because reservoir pressure in the field had declined to a point where it appeared that the gas reserves were depleted. "However," he reported, "the wells will eventually have to be plugged, and this could involve some additional expense to the working interest owners. . . ."

Meanwhile, in California, Rosenthal filed a variety of suits against Doris Day, alleging he was owed additional legal fees. She filed a counter complaint alleging that Rosenthal had been negligent in advising the Melchers in several areas, including oil and gas. She engaged the prestigious law firm of Irell and Manella to represent her and Melcher's separate estate. (The offices of Irell and Manella were, properly enough, at 1800 Avenue of the Stars.)

So Keplinger found assignments coming regularly through the mail on the "Doris Day" case. He was asked to make more complete examinations of wells and leases. He was asked to study financial reports, reports by other petroleum engineers hired by Rosenthal and others, and correspondence unearthed by Irell and Manella that seemed to have some bearing on the case.

It was not until March 4, 1974, that the case came to trial. It would last 100 days, involving a dozen lawyers. The Irell and Manella team was led by Robert Winslow, but the oil and gas proceedings of the trial were conducted by Thomas W. Johnson, Jr. Keplinger was his first witness, and the most important.

With Keplinger's testimony Johnson set out to prove that the oil and gas leases examined by Keplinger in 1969 had no value at that date; that the oil and gas leases in Calhoun County, Texas, had an "estimated net ultimate recovery value" during 1965 and 1966 substantially lower than the amount shown on the financial statement prepared during that period by Rosenthal or those acting for him; and that it was improper and inaccurate to rely on a 1962 geological study in the preparation of the 1965 and 1966 financial statements, particularly since production was decreasing during that period.

Almost immediately Johnson's adversaries were objecting to this and that, and particularly to what they called his "leading the witness." Keplinger was not accustomed to such rapid-fire verbosity. Judge Olson finally got in a word, saying, "The objection is overruled. A leading question is properly put to an expert witness as an exception to the usual rule excluding leading questions."

MR. NASATIR: Had I finished the objection, it would have been there has not been a proper foundation laid for this.

JUDGE OLSON: The objection is overruled.

MR. JOHNSON: I completed about half of the question, so I will rephrase it. Does this report accurately reflect your opinion that the Horton lease in McMullen County, Texas, was of no value on January 15, 1969?

MR. DUMMIT: Objection. There is no foundation for his opinion, at least at this time.

JUDGE OLSON: The objection is overruled.

MR. REGARDIE: It is immaterial also, Your Honor.

JUDGE OLSON: We might as well get things straight. Would you gentlemen decide who will handle the interrogation? I will not permit what has now occurred.

MR. DUMMIT: Very well, Your Honor; I will be handling it.

But shortly thereafter, Mr. Nasatir took over the questioning during a *voir dire* examination, a preliminary questioning to determine the competency of a witness. The questioning got far out of line, Judge Olson thought. "Just a minute, Mr. Keplinger," he said, halting an answer from the witness. "The Court is going to terminate this *voir dire*. It is obviously getting far beyond the appropriate scope of *voir dire*. It is in the nature of cross-examination, and the Court directs Counsel to desist. The direct examination shall continue."

And so it did, with Dummit objecting to almost every question Johnson asked Keplinger.

The 1965–1966 Rosenthal financial statement had estimated gross revenues for the Calhoun County leases at more than $2 million. Johnson asked Keplinger to assume that he had been asked to prepare a report on the properties back in 1966. "What would have been your dollar estimate of the total oil and gas reserves which would have been economically produced on the eight leases?"

Keplinger said, "I estimate that the value would approximate $32,000 as of that date."

Johnson wanted to know if it were Keplinger's opinion, then, that the estimated gross revenues were overstated by at least $1,950,000 in the Rosenthal financial statement.

"That is correct," Keplinger said.

Johnson asked him about wells on three other leases Keplinger had examined in Calhoun County—the Schultz, Lewis, and Sanford leases. "What did you determine regarding the Schultz lease?" Johnson asked.

"The Schultz lease had a permit granted for the drilling of the well, but there are no records that indicate the well was ever drilled," Keplinger replied.

The same was true in regard to the Lewis lease, Keplinger testified. As for the Sanford lease, there were no records that a drilling permit was ever granted. "And, of course, no records as to ever cutting the hole or drilling the well," Keplinger said.

This and similar testimony from Keplinger subjected him to a rough cross-examination from Dummit. The exchanges between the two men were exacerbated by Keplinger's refusal— or inability—to deliver yes or no answers. Time after time he told Dummit more than Dummit wanted to hear. On one occasion when Judge Olson intervened, Keplinger exclaimed, "I just wanted to be sure he [Dummit] understands."

JUDGE OLSON: Don't worry about that.

MR. DUMMIT: I'm not making the decision here.

JUDGE OLSON: Right. The small end of the funnel is here.

KEPLINGER: I'm sorry. Okay?

(LAUGHTER)

But a minute later Keplinger was reprimanded. "You are not to be concerned about the reason for the question, the thrust of it," Judge Olson told him. "That's for the Court to determine. If the question is not objected to, and the Court permits it to be asked, then you've got to answer it. If a person asks you what time of day it is, don't worry about why he is asking it or tell us how to make a clock. You just tell us what time it is. That's an example of what we are trying to get at. It will save us a tremendous amount of time."

DUMMIT: I might say, to alleviate your unrest a little, a further remark. I'm not trying to mislead you or trick you or anything. If I do it inadvertently, by not asking thorough enough questions, Mr. Johnson will then be able to clear that up; and anything you want to say, unless it is objectionable, you will be able to say at a later time.

KEPLINGER: It is not objectionable. I just wanted to include everything. I'm sorry, very sorry. It is important to this case.

JUDGE OLSON: All right.

MR. DUMMIT: Now, referring to the subject of the last five minute's conversation. . . .

And they were off again.

It was at the noon break that Judge Olson told Keplinger that Dummit had the right to examine the contents of his briefcase. He wanted Dummit to look at the papers during the lunch period to save time. Keplinger had papers referring to work in Abu Dhabi for Amerada Hess, among other things. "It could be a letter talking about the weather in Tulsa, and Counsel would be able to look at it," Judge Olson said.

But a little later, after some more conversation, Judge Olson said, "Mr. Keplinger, so that you will understand it, when we come back at 1:30 he would be entitled to look at this (papers and documents) piecemeal. I want to give him an hour and a half to look at this material independently."

When Court reconvened at 1:30, the cross-examination began. Because Potter Stanton, a former Keplinger associate, had written some of the letters seeking information during the examination of the leases and had done other work as well, Dummit questioned Keplinger unrelentingly about his meetings with Stanton, and then he faced Judge Olson.

MR. DUMMIT: At this time, Your Honor, I would like to make a motion to strike all the testimony that we had this morning that relates in any manner to exhibit 1128 [Marion County leases] on two grounds, the first being foundational ground, that it consists of a summary of completely unsubstantiated hearsay, and it is therefore hearsay itself; secondly, this witness is not qualified to authenticate this document in that he has just testified that there was a draft made by Mr. Stanton, no changes in the draft occurred, and that the words of 1128 were Mr. Stanton's and not his; furthermore, that Mr. Stanton had done a lot of the original research in this. All we have, in other words, is a summary of unsubstantiated hearsay, the gathering of which this witness participated in on a consulting basis. He did not gather the evidence, he did not prepare the summary, he merely consulted in its preparation, and therefore I would move to strike all of testimony this morning as it relates to exhibit 1128.

JUDGE OLSON: The motion is denied.

It had been a game try on Dummit's part. Judge Olson was hearing the case without a jury, and he had listened intently as Johnson had carefully, with plain and simple questions, extracted information from Keplinger. And in instances where he thought Johnson perhaps had asked too little about Keplinger's credentials and Dummit had questioned them, Judge Olson had asked questions himself to permit Keplinger to more fully explain the breadth of his experience.

Attorney Nasatir had another run at Keplinger. He asked many sharp questions, but Judge Olson finally wearied of what appeared to him to be a stalling tactic. And Nasatir apparently ran out of questions shortly thereafter.

Just as it seemed that the tiring day was at an end, up stood an attorney named Rhoads. "Your Honor, I have some questions," Rhoads said.

JUDGE OLSON: Just a minute, Mr. Rhoads. You will have to indicate your status in the case. The Court made it clear at the commencement of this morning's session that there should be a selection of counsel with respect to Mr. Rosenthal . . . and the selection was apparently Mr. Dummit, and that was confirmed by a further question by the Court. Your request to be permitted to cross-examine this witness is not permitted under any procedure the Court is aware of.

MR. RHOADS: Your Honor, I represent Mr. Rosenthal on the aspect of punitive damages on the defense case. The insurance company does not represent him to that extent, and I think it often occurs that an insured has his own Counsel on the issue of punitive damages even though he might have a carrier on the issue of negligence and even intentional tort.

JUDGE OLSON: There may be some area where that would be appropriate, but the request to examine this witness is denied, and the Court will have to struggle with that problem. It may have to require some kind of further election, because getting in innumerable bites at the apple would just be totally inappropriate.

(Keplinger probably felt like a well-bitten apple at this point.)

MR. RHOADS: Your Honor, I would point out that this witness gave an opinion concerning the evaluations contained in exhibit 259, and I intended to cross-examine him on that particular subject, which actually very well could be relative to the issues as to which I now assert I represent Mr. Rosenthal.

JUDGE OLSON: That is the scope of your desired cross-examination?

MR. RHOADS: Yes, Your Honor.

JUDGE OLSON: The request is denied.

☐

It had been mentioned several times during Keplinger's day-long testimony that he had to catch a plane at 5:30 p. m. Keplinger himself mentioned it at one point, and Johnson had used it in objecting to Dummit and Nasatir's lengthy cross-examination.

As Keplinger was being dismissed from the witness stand, Judge Olson had a word with one and all.

JUDGE OLSON: One last comment. Mr. Johnson, the Court has in mind your reference with respect to the previous witness being from out of town, and your request to terminate the cross-examination, which was denied. Part of the time today was consumed by the fact that the witness courteously, interestingly, and with a nice, soft Texas accent, answered frequently in much greater detail than the question called for. If we had nothing else to do, it would be interesting; it would be an interesting, you know, exercise in something or the other. . . ."

Several days after his testimony, Keplinger received a letter from Johnson. One paragraph said: "I want you to know that we were especially pleased with your testimony, and the way that you withstood a rather rough cross-examination. You might be interested to know that Jim Ramsey and Wendall Cook [consultants who followed Keplinger to the witness stand] were not subjected to a very lengthy cross-examination. We feel that their job was made easier by your testimony. . . ."

The trial lasted for several more weeks. At its conclusion, Judge Olson delivered an oral opinion instead of a formal, written one. "For this Court to write a formal opinion would be an exercise in futility, or an ego trip," he said, which was the first indication of how open-and-shut he considered the case against Jerome Rosenthal to be.

Total damages awarded Doris Day and the Melcher estate amounted to $22,835,646, the largest amount ever awarded in a civil suit in California to that time.

Of that total, $5,589,000 plus interest of $1,956,150 was awarded for damages in oil and gas ventures.

No longer financially strapped, Doris Day broke into tears and told reporters, "I am healthy, and I am rich!"

In Tulsa, Keplinger read the story of the verdict in the newspaper. He tore out the story and scribbled a note to his secretary on top of it: "Anna, please write them for the papers I left out there."

CHAPTER NINE

I t will be recalled that Keplinger began writing scientific papers while he was still attending the University of Tulsa. He was to continue this practice throughout his career. He wrote for journals devoted to the various engineering disciplines, and he wrote for the popular oil and gas magazines. Many of the presentations he made before the societies to which he belonged were later printed in the appropriate journals. And he liked nothing more than writing a piece for a Latin American periodical where his proficency in Spanish was appreciated. Later, he would write for French publications as well.

He was fortunate in his early years at Shell that his reports to various levels of management had to pass under the eagle eye of Ted Swigart, an engineer who was to become president of Shell Pipeline Company. In his youth Swigart had been employed in a government agency where stricter attention was paid to a report's construction and punctuation than to its contents, or so it had seemed to him. Time after time he returned Keplinger's reports for polishing before allowing them to continue up the ladder to the proper level. Keplinger grumbled at Swigart's constant demands for revising, but he certainly profited from it.

Later his writings would pass under the sharp pencil of Anna Brown, his trim, attractive secretary. She edited his papers, just as she edited almost every report that left the office. Like Swigart, she was able to correct his obvious errors—but no one could teach him how to spell. And, thankfully, no one ever attempted to curb the restless poetic spirit that oftimes bubbled in his prose.

Many Keplinger articles dealt with secondary recovery efforts as the state of the art improved. He was considered an expert in the field, and he wrote with authority. And, as an old friend remarked, "Henry writes so damned much about unitization you'd almost think he invented it." Indeed, while he was acting as consultant on the unitization of Redwater Field in Canada, he delivered a paper on the subject at a conference in Banff in which he recited the long and involved history of the process. The paper was published later by the *Western Oil Examiner* in Calgary, and could stand today as a textbook on unitization.

But it was in his work books where his love of language was most evident. He was not a demonstrative man. "Some mornings I feel like I just met him for the first time," Anna Brown told a reporter in explaining that Keplinger was a very private person. Keplinger was not one for shooting the breeze in the office; he allowed no political prattle and never brought his personal life to town with him. But occasionally he would write long memos in his work book—memos no one else ever saw— and in them the inner man often was revealed.

He wrote one such memo while on a trip to Europe for a long-time client, Morris Mizel. Mizel was a Tulsa wildcatter who operated as an individual in Oklahoma, Kansas, Texas, New Mexico, Canada, Alaska, Guatemala, and Australia and actively sought concessions elsewhere in the world. He was aggressive and undertook deals of large size and scope. Keplinger evaluated these far-flung leases for Mizel and worked for him in other capacities as well. In this instance, Mizel sent Keplinger to St. Moritz, Switzerland, to discuss a Turkish concession with Tex Feldman, a Beverly Hills, California, oilman.

He arrived in Zurich late at night during a snow storm:

> I was driven to town in an excellent bus, 75¢ trip. The air terminal was very modern and the roads perfect with overpasses even for pedestrians. The highway was completely lighted with a row of center lights. All the buildings so modern.
>
> Arrived in town a little after midnight. Got to bed about 3:15 a.m. after making arrangements to go to St. Moritz at

6:55 a. m. by train. It is a 4½ hour trip along the lakes and through mountain passes. Had a continental breakfast in my room.

Everybody speaks German in these parts, but I was able to get a French radio station with the news. Went to the railroad station, which was very busy—lots of people going skiing. Also many commuter trains bringing people to work for 7 o'clock starting. It is 15° below zero, and everybody is bundled up.

The train trip carries us through some beautiful sawtooth mountain country. The sun came out about 8:30 a. m., and it makes the snow glisten on the sides of the mountains.

We changed trains at Chur, and I bought a chocolate bar to give me strength. I wish I had reviewed a little more on the German [phonograph] records, as all the announcements are in German. While I have become proficient in French, I have grown rusty in German.

The trains operate on narrow gauge tracks. All carry lots of baggage cars. The passenger cars are heated by electricity. The inside temperature is 13° Centrigrade, so I have my coat on.

Arrived at St. Moritz at 12:15 p. m. and went directly to the Hotel Palace, operated by Badrutt. It is the winter watering ground of nobility—English, French, and German. German industrialist Gunter Sachs was in the lobby with special (Lynx) coat. Danish prince with 12 children, etc.

Had lunch with Mr. and Mrs. Feldman in Petite Room overlooking the lake. They were having auto races on the lake, and just below was a skating rink where several people were doing beautiful figure skating.

The lunch was marvelous, 2 hrs, with black raspberry juice, Hungarian Goulash, vegetables, wonderful green salad, special soft cheese, wine, fruit & nuts plus special cookies and espresso coffee.

After lunch we visited about Turkey and called Paris to visit with Jerry Medisco. Feldman agreed to file for Mizel, and I called Mizel in Tulsa. Prepared a brief memorandum letter on deal which Feldman signed.

Before dinner I took a walk around town and looked at ski lift. It was cold and everyone was in ski clothes. The town is up and down. They had several sleigh horse-drawn taxis which looked like the equipment used by the

Russian Czars. Visited several of the shops, mostly ski or heavy winter clothes or watches.

Had dinner with Sir Charles Oppenheimer of London, Ernest Kantzel of Detroit (friend of Waddy Wadsworth), and Nichols McLouwin of London (Greek ship owner), and the Feldmans. The meal was extra. Main course was Wellington Roast (lots of beef wrapped in a paste dressing and baked), a specialty of the house, wonderful cold soup, salad, vegetables, cheese, sherbet of Cointreau, fresh fruit, coffee and brandy. Lots of cookies, which I did not eat.

After dinner we retired to the Main Hall for more brandy and coffee, and to see the celebrities walk by. I was seated by Mrs. Pat McLouwin for dinner. She is an English girl married to a Greek. She has been coming to St. Moritz both winter and summer since she was a child and is an excellent skier. She was most pleasant, and had skied every day, up at 7:30 a. m. and up to the top by 10:00 a. m. to be one of the first. Then back to the top for lunch at the private club where they can have the same guest only once each season. I was invited up for lunch if I would stay another day.

My other dinner partner was Sir Charles' "fiancay," Diana, who had terrible stories to tell of her experiences in Japan prison camps during the war. She was visiting her parents in Shanghai and was taken prisoner.

After dinner I sat with Mrs. Jane Feldman and learned about her life and children. A native Californian. Stanford graduate, tennis enthusiast etc. Had a brief visit with Rose Marie Kantzel who knew the Wadsworths. She was a 'Gabor' type, wordly, 5 languages, and a 'ski bug.'

In the casino they had an orchestra from Rio, and the younger ones were doing the Bossa Nova.

After dinner I rewrote agreement for Mizel-Feldman and got Feldman to sign just before midnight. In bed at 12:30 after packing.

Up at 6:15 a.m. Got 7:30 train for Zurich. Breakfast on train with M. Menzie of Philip Hill, Higginson of Johannesburg. He is also director of General Mining (Diamonds). British to the core and most interesting. War record with John Stanley etc. His mining company is interested in getting diamonds from the sea. He will work in Zurich for a day and then back to Johannesburg.

We changed trains at Kurd, about half way between Zurich and St. Moritz. It is a little railroad town hemmed in by mountains on all sides and completely snowed in at this time of the year. All the house roofs have 1 foot or more snow, and all the chimneys have wisps of smoke floating upward for the wind to play with. Every once in a while along the route you see old fortified areas which have complete provisions for safety.

Visited with an English family on the train from the Island of Jersey. Arrived at Zurich at 12:45 p.m. Visited around city, but it was too cold to enjoy much, was −4° F with fog. The airport was closed for four hours. Had lunch at the airport, excellent ham, and left for Paris at 3:35 p.m. by Air France plane 638, a Caravelle, clear, excellent flight over Switzerland. Everything frozen on the ground. Lake Zurich is frozen for the first time in 50 years. While flying over Basel, we had a magnificent view of the Swiss Alps to the south. The structure trends in the north of Switzerland were easily discerned from the air. Most of the people on the plane are French now.

After a most pleasant trip we landed at Paris at 5:00 p. m. and Jerry Medisco met me and took me to Hotel de Crillon and then to his office. Worked until 8:30 p.m. and went to hear Yves Montand sing at the Etoile Theatre. He was at his best with all the new French popular songs. Then to Lido, magnificent show, revues, acts, ice skating, western act, dog tricks, tumblers, and girls with and without clothes.

It is terribly cold. Got a copy of *Le Monde* to read, and *La Figaro*.

Went to Feldman's office to go over Turkey and to determine if they were still "a permittee."

They were not. Called Mr. Feldman. He would like to go ahead but it would be more expense.

Lunch at Restaurant Bigart. Visited J. N. Legrand with SNPA, also Mr. Tavernier with Forasal.

Talked to Mr. Feldman and Mr. Mizel. Cancelled trip to Turkey and made arrangements to go to New York.

It was snowing and very cold when I left the office at 7 p.m. I walked to the hotel. Tried to call Ward Dunn with Phillips, out. Took hot bath and went to bed. Woke up at 9:30 p.m. and took a walk up to the Opera House. Nobody on the streets. Very cold. Came back to hotel and had dinner and wrote cards.

> Up at 7:00 and went immediately to the airport. Plane left on time at 10:00. Flying at 28,000 feet. Expect N. Y. at 12:00.
>
> Had an uneventful trip and slept about ½ of the way. The food was not as good as on TWA going over, which is unusual. Read some French and collected some French newspapers. Landed at 11:50 a. m. . . .

What did he have in mind when he wrote such memos that would molder in his work books . . . memos that would go unread unless discovered by chance? He often described parts of his journeys in the letters he wrote to Louise and his children, but not in this detail, and not so objectively. And the business notes in these strange memos could have been jotted down on a single page.

In the memo just quoted, before he arrived in Zurich, he had written:

> The Common Market is the one great continuing news story of this period. The economic health, development of undeveloped nations, and the political stability of the future is dependent upon the eventual solution. Brussels, as the center and present seat of the Common Market, is making the headlines every day, in the newspapers and radio and TV.
>
> DeGaulle's infuriating and didactic speeches have brought forth strong replies from Belgium and its foreign minister, Paul-Henri Spaak, Dutch foreign minister Joseph Luns, and the other members of the Common Market. Even the father of the Common Market, Jean Monnet, takes issue with DeGaulle. . . .

These words were not written by a man who put them on paper so that he would remember random thoughts to pass on to someone else. They are, it seems, the reflections of a man who wrote them simply because he felt compelled to write them. Indeed, they appear to have been written by a man who longed desperately for time away from his profession so that he might at leisure present his thoughts and attitudes to be read by others.

As time passed, he would now and then write fully—in a

separate notebook—about some event in which he had participated. At times he would have these essays typewritten and would pass copies to his employees. He was gaining confidence in his skill, and like a small boy testing a strange stream with his big toe, he was looking for a reaction to his literary work.

One of his reports described his visit to the famed Bohemian Grove as a guest of a friend, Jack Sembower:

Bohemian Grove lies 80 miles north of San Francisco and was organized in 1872 by a group of San Francisco newspapermen to raise the cultural status of the Bay Area. Membership is limited to 1,000. Of these, 100 must be actively engaged in the arts, such as actors and musicians (engaged in daily efforts). Fifteen honorary life memberships are permitted. Three hundred fifty members are selected on an annual basis. Theoretically, at least all of these are able to contribute something worthwhile in the way of entertainment. This brings in a fabulous group of actors and musicians, and the list reads like a Who's Who of the American Stage and Television. Seven hundred fifty out-of-town men are admitted to membership.

Shortly after the club was organized, the great lumberman Meeker sold much of what is now the present Grove to the Bohemians for a permanent preservation, which covers better than 3,000 acres and is surrounded by mountains on either side of the valley and extends to the banks of the Russian River a few miles above the point where it enters the Pacific Ocean.

The members and guests are housed in approximately 120 camps. The accommodations of the camps vary from moderate setups to rather pretentious and even luxurious affairs. During the two-week retreat in the woodland fantasy, no women, automobiles, radios, or televisions are permitted. Transportation is provided by roofless buses. Cameras may be used except at the swimming hole. No one is called to the telephone, but telephone calls are posted on a blackboard.

Parking is a problem. Everyone must check in and out during their stay at the Grove. My experience was that the baggage was at the camp before I arrived.

Close to the entrance, on the floor of the valley, is an

infirmary. From the entrance, we walked to SPOT Camp and admired the great Redwood trees some 10 feet in diameter, over 2,000 years old, and more than 200 feet tall. As we walked along, the sun would occasionally break through the trees along the trail. Some 1,500 men were checking in and renewing old friendships.

We arrived at SPOT and I found a beautiful courtyard with an open fireplace and an enclosed area with a fine piano. Above the courtyard on the side of the hill are camp facilities for 15 people, with excellent beds, electric blankets, and plenty of toilet facilities. All of the members and guests were wonderful and made you feel at home immediately. Our Captain Morris Cox radiated friendship. Small discussion groups settled the problems of the world, including Nixon's recent China statements. There were many, many good jokes.

We could visit other camps by walking or using roofless buses especially designed and built by Chrysler Corporation for the use of the Grove, but now supplemented with some General Motors models.

At about 8:00 p. m., some 1,100 men had steak dinners in a beautiful Redwood clearing, and we finished up with apple pie a la mode. The dining hall was lighted with hundreds of gas jets which made a most cheery sight as they flickered through the Redwoods. At 9:15 there was an excellent campfire program.

The campfire circle had an open air fire burning in the center of the circle and we seated ourselves on freshly strewn Redwood leaves to enjoy a musical and literary program, and observe the smoke drifting up through the Redwoods. This night, with lots of wit during the ceremony of the spreading of the leaves, we were told some of the mysteries of the Grove. The symbol is the owl—because of his traditional wisdom; and the motto, 'weaving spiders come not here,' means business must be forgotten within the Grove and no financial deals transacted. As I wrote in camp some 150 feet above the floor of the valley, the smoke from the campfire in the patio drifted up into my cabin.

After the evening program, we roamed the camp grounds visiting friends who had asked us to come by. Each camp had its own individual charm. Finally to bed at midnight, my first day at the Grove.

We were up early and had lots of discussion in the camp before going to breakfast. In a nearby camp, the musical instruments were going—in our own camp, a fine piano player knocked off the chords from classical to jazz. I had a chance to read the morning paper. . . . still about China and Nixon's plans to visit.

After breakfast we went for mail and visited the museum. Then a lovely luncheon in our camp. We visited more camps and friends until dinner time. Our dinner of roast beef, under the Redwoods, was followed by speeches from the new and old guard. They emphasized friendship and fun. Dinner over, the entire assemblage walked to the lake behind a torch procession which came down the hillside for the official opening of the two-week session in an impressive ceremony known as the 'Cremation of Care,' with fireworks and pageantry which symbolically reminded all that the two weeks in the Grove were to be free of all worldly problems. At the end of the lake is a stage backed by a huge moss-covered masonry owl. Before the owl, around the funeral pyre, the Chief High Priest and his helpers were gathered. Amidst funeral chants and solemn music, a boat appeared around a bend, poled by a lone boatman. We were all on the shore of the lake. In the boat was a casket bearing 'Care.' Just as the boat reached the funeral pyre, a brilliant spotlight was flashed on a huge Redwood totem pole on the hillside which was painted pure white. From its top came weird laughter and a mocking voice cried: 'Fools! Fools! Fools! To think you can banish Care! Burn me tonight if you will. On Monday morning I will be waiting for you as usual in the Market Place!'

Everything about the pageant was perfect and it ended with a fireworks display. Then we visited several camps. All had a cheery wood fire and lots of fellowship—some had a fine musical program. We finished the evening at a camp which was serving a lot of wonderful hot soup.

My routine was to scan the morning paper, which was the San Francisco *Examiner*. The main highlight was Red China. Nixon's proposed trip was discussed, editorialized, and politicalized from the front page to the back page. The agreement for President Nixon to visit Peking looms as a major milestone in the geopolitics of the 20th Century, with ramifications extending far beyond the war in Southeast Asia. The invitation should ease tension in Asia and enhance prospects for a Vietnam peace settlement. Everyone in our camp was in favor of the plans to diplomatically link up with China. It was rumored that Dr. Henry Kissinger, who conferred with China's Premier Chou En Lai, might show up at the Grove because he had been there last year.

We took a Sunday walk around the rim of the Grove. It was up-and-down and I was glad to get back to camp in one piece—but it was a wonderful communion with nature to walk among trees which have been growing for over 2,000 years. The authorities have adopted the name Sequoyah as the scientific name to cover all the genera of Redwoods. Before lunch, we strolled over to the lake to rest in the sunlight and listened to beautiful organ music.

The afternoon program was a running dialogue by Astronaut Charles Conrad about his trip to the moon. He was a great and descriptive storyteller and made you become a part of the trip from the time a lightning bolt almost destroyed the flight to his walking on the moon.

The Sunday evening program was musical and storytelling. A great jazz band and lots of singing by individuals and a great chorus with Grove members doing the entire program, as always, in the Grove. This ended my third day at the Grove, and I enjoyed every minute.

Come Monday morning I packed my bags and with great regret left the Grove and all its fun to go to San Francisco. I will long remember the jokes and the many discussions about health, youth, politics and, of course, the energy crisis in the United States. . . .

And then he had written:

> Why did the Russians leave it
> And go back to the snow?
> This little bit of Heaven
> I'll never, never know.

□

An interviewer once asked Keplinger what prompted him to learn the French language.

Said Keplinger candidly, "We had no way to get into France and make a dime unless somebody could speak French."

So he joined the *Alliance Française* of Tulsa, and became its president. And every Friday afternoon he left his office to visit a French tutor.

The tutor was an elderly lady, so delightful and solicitous of Keplinger's studies that Louise and the children jokingly called her "Daddy's girlfriend." He learned to speak, read, and write French as rapidly as he had learned Spanish. The tutor told him he spoke French with a Spanish accent, not an English one.

Keplinger's knowledge of French paid off, and not just in France. One of his clients, National Cooperative Refiners Association, often sent him on European jaunts, and Louise occasionally accompanied him. One trip took him to Brussels, where he tried in vain to obtain some needed information from Belgian officials.

"Henry found out that the information might be in a Brussels library," Louise said. "It was, but it was in French. Of course, that presented no problem to Henry. He learned all that he needed to know right there in the library." She added: "Some of his friends would kid him about wanting to learn foreign languages, but Henry knew what he was doing."

Other languages also came easily to him, though he made no long-term study of them as he had with Spanish and French. "On a trip to Japan," Louise said, "he worked very hard on his Japanese. The people at the American embassy laughed at him and said it was impossible to learn the language without a lot of time and study. But Henry plugged right on, and he managed to learn enough in a short time to get along."

One of Keplinger's foreign assignments took him to Yugoslavia, where he evaluated the country's gas reserves for the government. Eight months after he completed the assignment, he and Louise were at a party in Oklahoma City where some of the Yugoslavian officials Keplinger had met were in attendance.

"Your husband simply amazed everyone," one official told Louise. "No one who comes to our country knows enough of the language to converse, but Henry got out among the people of the towns and talked with them. They were impressed that he learned to talk with them in such a short time."

Keplinger learned to speak Italian while estimating the gas

reserves in Italy's Po Valley. "Mattei helped me with my Italian," he once told an interviewer. "Mattei" was Enrico Mattei, brilliant, mercurial chief of AGIP, the Italian National Oil Company. The two men had lunch together on several occasions. The meeting must have been interesting, for words flew from Mattei's mouth like bullets from a machine gun. While talking with a foreigner, he was apt to carry on a separate conversation with a compatriot while his words to the foreigner were being translated.

Keplinger flirted with Portuguese and other languages as well, but Spanish remained his favorite. He addressed groups in several Latin American countries in their own tongue, but nowhere did he make such a hit as he did in Spain.

He and Louise were in Madrid for the initial meeting of a Spanish engineering society. Keplinger had been invited to speak at the meeting, and he and Louise took the opportunity to spend a couple of weeks in the city before the big event.

They stayed at the Ritz Hotel, and Keplinger, as was his custom, honed his Spanish on bellboys, maids, and elevator operators. Every employee and guest in the hotel soon was aware that the American was going to address the engineering meeting in Spanish. And all were hoping he would be a big success.

Two days before the meeting, Keplinger lost his voice. The sad news swept through the hotel. Scores of anxious eyes watched the hotel doctor as he went to and from the Keplinger living quarters. Would the doctor be able to restore the American's voice before the meeting?

Confident that the doctor would succeed, Louise went on a sightseeing trip the day of the speech. When she returned to the hotel, the elevator operator said excitedly, "He did it! He made the speech!"

There must have been some question around the hotel for several days as to who was the bigger hero, Keplinger or the doctor.

☐

Enroute to Madrid to deliver the paper, Keplinger and Louise had stopped at Kassal, Germany, to visit with some German investors who wanted Keplinger to review the geology of offshore Turkey for them. They stayed in a beautiful hotel on the edge of a park while Keplinger did his work. But Keplinger was eager to make a side trip to Gottingen where he had spent so many wonderful days as a university student.

The city had been bombed by the British during World War II because it had become a war materials manufacturing center. Keplinger had received letters even during the war from friends he had made as a student.

Once in Gottingen, Keplinger took Louise on a tour of the university, showed her where he had lived, and took her to dinner at the Rathskeller on Weenderstrasse, in the heart of town, where he had eaten regularly as a student. "Well," he told Louise, "they're still serving the same kind of food in the same kind of way—and it's still good." He was thrilled.

They drove out of Gottingen in a hurry because they had to catch a plane in Frankfurt. As they rode along, Louise spied on the map the little town of Spang, not far off the road they were traveling. The community was her ancestral home. But the plane in Frankfurt, they knew, wouldn't wait, so they reluctantly bypassed Spang, vowing to return at another time. They never did.

⬜

Louise accompanied Keplinger on business trips to most of the European and South American countries, but perhaps their most memorable journey together was to Kenya and South Africa. The South African government wanted Keplinger to study the country's oil potential and examine its coal fields. Kenya was an exciting stopover.

In Nairobi, the Kenyan capital, Keplinger visited with government officials and engineers—a courtesy call—and took Louise to the Rotary Club meeting. She always went with him to Rotary meetings when they were abroad. And as he always did when he was in a foreign city, Keplinger presented the Nairobi Rotarians with a Tulsa Rotary Club flag, which they graciously accepted.

The government offices Keplinger was visiting were in a big building of modern design. Directly across the street from it was an area packed with tribal tents. Looking at the tents from a window in the building, Louise thought, "There they are, and here is where they're trying to get."

The highlight of the Kenya stay was a night at the Treetops Outspan, a hotel built on stilts in the town of Nyeri. In clear view of the hotel was a great waterhole lighted by giant flood-lights. And at night, the animals came in groups, unmindful of spying eyes from the hotel, to slake their thirsts. The Keplingers stayed up all night to watch them.

In a letter to his associates in Tulsa, written from the Treetops Outspan, Keplinger wrote: "Last night we saw 26 elephants, 19 rhinos, 245 Cape Buffalos, 4 Giant Forest Hogs, 121 Warthogs, 2 Hyenas, 44 Waterbucks, Bongo Antelopes, Baboons, Mongoose etc. The day before, Zebras, Giraffes, Lions, Leopards, Hippos, and you name it. A memorable stay in the atmosphere of the African World. . . ." And then there followed a list of things to be done in Tulsa—things he wanted attended to right away.

Louise later told her friends, "I stayed up with Henry, but I didn't do any counting. But you know Henry. He never misses anything."

They traveled on to Johannesburg, South Africa, the city that virtually sits on top of the richest gold deposits in the world. It was in this city that Keplinger did most of his work for the government. The couple was royally entertained in the fascinating city. But both were anxious to visit Cape Town, and they took off several days to make the trip.

Cape Town, about a thousand miles southwest of Johannesburg, is the southern extremity of Africa, and it is there that the Atlantic and Indian oceans wash against the fabled Cape of Good Hope. The city began in 1652 as a ship-victualing station established by the Dutch pioneer Jan van Riebeeck. To sailors, the rugged peninsula itself was known as the "Cape of Storms."

Keplinger sent a postcard bearing a picture of the Cape to his office associates. "Visiting this area. Our study covers an area of 85,000 square kilometers to the east of here. It was great seeing the surface outcrops on-shore."

They both climbed a series of steps from the beach to the top of the Cape, a wearying jaunt, but Louise called it "one of the most thrilling moments of my life."

CHAPTER TEN

At long last the wanderer got to Russia. He was amazed at what he found there—a secondary recovery technique he had not seen before or even read about in all his years of working in the far corners of the earth.

In Germany, while a student at Gottingen University, he had visited an oil mine in the Wietze field, it will be recalled. There, a 300-foot shaft had been sunk into a shallow oil sand. The sand was mined and carried by a continuous conveyor to the surface where it was washed with hot water in a cylindrical tank with a cone-shaped bottom.

But in northern Russia, not far from the Siberian border, he found oil mining being conducted on an enormous scale and at a most sophisticated level.

The Russian assignment came from the Resource Sciences Corporation of Tulsa. Jack Peterson, then vice president of the corporation's technology division, was an old Russian hand. He had been behind the Iron Curtain many times for RSC and also for a former employer. He spoke the language well and was at ease dealing with the Russians. His group was responsible for selling Russian technology in the U. S. RSC was an engineering company, not a marketing company, but its officers reasoned that RSC would be the logical choice to engineer a project for a customer to whom it sold the Russian technology.

Buying and selling of technology was conducted for the Russians by Licensintorg, a division of the Ministry of Foreign Trade. Peterson learned that Licensintorg was interested in marketing the technology of thermal shaft mining. It sounded as if it had interesting possibilities, so RSC invited a team of

Russian experts to Tulsa to explain it. And Keplinger was engaged to evaluate it.

The Russian interpreter was Dr. Yusuf G. Mamedov, who had studied petroleum engineering at the University of Kansas. He had read so much about Keplinger and had heard so much about him that he almost was in awe of the American consultant. And Keplinger was surprised to learn that the other Russians also were familiar with his work. They had read all of his scientific papers and had discussed him in their own scientific circles.

After several meetings, RSC officials decided that the technology must be seen. Peterson, Keplinger, and Peterson's assistant, Peter Richardson, flew to Moscow in late October. Their final destination was the Ukhta field, a vast shallow reservoir that had been discovered during World War II and exploited by slave labor.

The oil in the huge reservoir was heavy and had proved difficult to produce by primary recovery methods. Steam injection wells had been drilled, but this method also proved unsatisfactory. Then a Russian scientist had come up with an unusual idea, and it was the fruition of this idea that Keplinger and his companions were to examine.

Booted, wearing a mountain of clothes and a steel hat, Keplinger was taken to the Yarega mine. The men got into an elevator and dropped 600 feet into the frozen earth in the blink of an eye.

The Russians had constructed a maze of galleries, both horizontal and vertical. The group walked along the horizontal galleries as if they were strolling along the corridors of a great building.

But every so often they would come upon a big valve jutting out of the wall like a spigot on a wine cask. One of the Russians would turn the spigot, and oil would run out and fall to a trough on the gallery floor. Another spigot might release water; another, nothing; the next, oil. Oil and water ran down the trough to a large separator tank.

The spigots were connected to horizontal perforated pipes which reached back into the reservoir. Steam was injected from above. The heated oil made its way into the perforated pipes to be released at the spigots. When the oil and water were separated, the oil was pumped to the surface; the water was recycled back to the power plant to be converted to steam and again injected in the reservoir.

The injection wells were not drilled from the earth's surface but from galleries above the production galleries. The Russians had developed a drilling rig for this purpose, and its design was included in the technology package.

Keplinger was given access to most of the information on the project. The Russians held back, however, on the data for what they considered a unique steam injecting process. "After all," said Dr. Mamedov with a smile, "no one had paid us any money yet. We want you to have just enough to be able to sell the technology."

On the surface the group was ushered into a shower so that every drop of oil on their boots, clothing and bodies could be recovered. The droplets went into a sump. The Russians, it appeared, prided themselves on their high recovery efficiency.

The party was lodged in an old hotel which had been converted from a schoolhouse. The Russians all raced through the snow to a sauna, where they would steam and beat each other with birch twigs before rolling in the snow like bear cubs. Also, a lot of vodka was consumed in the process.

Keplinger had recently undergone surgery which made him walk gingerly. So he passed on the sauna games and stuck to the vodka. The women of the small town got together and provided a feast for the party and the hosts. "It was a fantastic meal," Peterson recalled. "All of the native dishes were served, so much that we couldn't make a dent in it. And later there was a huge cake, tiered like a wedding cake."

The party returned to Moscow. The Russians insisted that the Americans go with them to a plant where other Russian scientists had devised a process to produce champagne in a very short period of time. Keplinger found it hilarious when his hosts said they had sold the process to the French!

There was a sipping room where the party sampled seven varieties of champagne. "Most of it was awful," Peterson said, "but one, very dry, was excellent. It could compare favorably with the best French and American champagnes."

The party left the plant by means of a long ramp. Keplinger stepped on a spot of oil and slipped, falling right on his backside. The Russians were frightened; obviously they had not kept the area spotless. They rushed to Keplinger, but he got to his feet with a smile on his face. "I'm all right," he said, laughing, "but I want you to know that it wasn't too much champagne that made me slip!"

☐

Keplinger had been impressed by the Russian technology at work, but he was not fully convinced it could be applied economically in the United States.

The Ukhta field, while shallow, was extensive; most heavy oil reservoirs in the U. S. are small. At Ukhta, the producing zone was "competent"—when the reservoir was heated the oil flowed out and the rock stayed put. In North American heavy oil fields, oil and sand both flow when the reservoir is heated.

So Keplinger realized that it would be difficult, if not impossible, to find a reservoir like the one at Ukhta—one with enough oil and good rock to make application of the technology a commercial success.

Some companies with leases in the Athabasca Tar Sands showed interest in the technology, but apparently they never could convince themselves that it was economically feasible.

☐

Some of the Russian scientists had returned to Tulsa with Keplinger, Peterson, and Richardson, and the Keplingers invited them to their home for a party. In attendance were many of the Keplingers' Tulsa friends. The Russians had a fine time and were impressed with the home, the furnishings, and particularly the mementos of Keplinger's travels.

The Keplinger home, at 2213 East 27th Street, was the scene of many parties at which foreigners were the guests of honor. Keplinger had made friends with foreign students while he was attending the University of Tulsa, and he continued in later years to seek out foreign students at the school and help them in any way he could. Many of these young men returned to Tulsa years later as representatives of their governments, and they always seemed to find their way to the Keplinger residence.

When Keplinger had first gone into business for himself,

Louise, of course, was left alone at home much of the time. She was bewildered by this pattern of life, unaccustomed to making the daily decisions necessary to maintaining a household and one's own mental equilibrium. Ferdinand Spang, her father, in Tulsa for a visit, took her in tow.

"Louise," he said, "this is going to be a hard venture. You're going to have to make up your mind to go ahead and make a life for yourself. You're not going to sit home every day with the children. You be sure you have babysitters so you'll be free to go and come. Get involved in things. Be free to entertain. By making a life for yourself . . . that's how you can help Henry the most."

Louise would say later, "That's when I grew up."

She began volunteer work ("Mix," Keplinger's grandmother had told him when she bought him the trumpet), becoming active in the Tulsa Boys Home, the Tulsa Geological and Geophysical Society Auxiliary, the AIME Wives Club, and of course the Rotary Anns. She already was active in the Daughters of the American Revolution.

And she threw open the door to her home. "We always thought we should have people in our home," she once said. "We wanted them to see how we lived, and we wanted them to meet our children. We may have had just a cocktail or two, and then gone out for dinner, but we thought it important that they visit our home first. Then, of course, we often had dinner with them at home."

Many times, however, she opened the front door to see Keplinger and a stranger on her porch. Anna Brown, Keplinger's efficient secretary, vividly recalled an occasion when Keplinger invited Joseph Laport, a Brazilian oilman, to dinner. Keplinger and Laport were visiting in Keplinger's office shortly after noon when the invitation was tendered. Laport left the office to make some other calls, planning to rejoin Keplinger at five o'clock.

As the day wore on, Anna fretted because Keplinger had not phoned Louise to announce that there would be a guest for dinner. Finally, she suggested to Keplinger that he call home. "You don't want to surprise Louise," Anna said.

Keplinger shook his head. "It's not necessary to call her, Anna. Whatever we have, we'll share with Mr. Laport. We'll just set one more place at the table. When you invite someone to your home, all you have to do is to be sure that the kids' faces and hands are clean."

Said Louise: "I didn't mind the last-minute dinners. Henry was gone so much that I was glad he was at home under any condition."

The Henry Keplinger at the office and the social Henry Keplinger were vastly different human beings. It would not be correct to say that he did not allow levity among his employees, but he certainly didn't encourage it. If someone had a pleasant little story to tell, Keplinger had a way of letting the person know that one telling was enough.

And he abhorred office politics and gossip. If A came to Keplinger with a complaint about B, Keplinger would stop him before he could utter it. "Anna," he would say, "tell B to come in here." When B arrived, Keplinger would say to A, "Now let's hear it." Such confrontations were rare, but so was office backbiting.

The fact was, he came to the office to work, and he expected everyone else to have the same attitude. Even in his correspondence, he never dictated a personal letter to Anna Brown. He wrote personal letters elsewhere. And there was not an office supply of alcoholic beverages for his clients, casual visitors, or employees who thought a toast was in order because of a business *coup*. Celebrating was for after-hours.

But at parties he did not frown on alcohol, and he was quick to enter into the spirit of the occasion. At the 1975 Christmas party for employees, clients, and friends of the Houston office, held in Vargo's Restaurant, a stranger appeared. All eyes were on him as he moved down the stairway from the restaurant level to the crowded cocktail lounge. Most of the men at the party were in formal dress, the remainder in dark business suits. But the stranger wore dark glasses, and he was dressed in a burnoose, the traditional hooded cloak worn by Arabs and Moors.

The Arab oil embargo of 1973 was still fresh in everybody's memory, and OPEC had become as familiar an acronym as M*A*S*H. Was this stranger a representative of that imposing organization in the Middle East?

He moved past several knots of people, arriving at last at a group that included Betty Messersmith, young Kep's charming and efficient secretary. The stranger bowed before her, took her hand, and spoke in Arabic as if he were introducing himself. Startled, Betty took a step backward. Then a smile came on her face. "Mr. Keplinger!" she cried, and the smile broke into laughter.

Keplinger it was. He moved on, going from group to group until his identity was no longer a secret. Said an old friend: "You could always expect Henry to do the unexpected."

And some wonderful parties of all kinds were held in the Keplinger' comfortable home. Indeed, the lower floor of the two-story house seemed to have been designed with parties in mind.

On September 27, 1961, Mary Lou Bulloch described the house in a story in the Tulsa *Tribune*:

> Inside, the large entry hall is really an introduction to the entire living area. Standing in the foyer, one can see through the wide archways into the other rooms—the formal dining room on the left, the living room on the right, and the family room beyond. . . .
>
> The entry hall attains its look of spaciousness partly because of the extra wide openings into its connecting rooms, and partly because it is really of generous size itself. . . .

Of the clubroom, she wrote:

> No one could stay in this room long without guessing the international ties of this family. For on the game table is a brass scale with a bouquet of flowers balancing a bowl of foreign coins. In the open cabinet nearby are many items from other lands—a beautifully carved wooden wine jug from Yugoslavia, a dainty caravelle, or fish-boat, which Karen purchased last summer in Portugal, and graceful statuettes of animals fashioned of rhino tusks, from India. . . .
>
> On tables throughout the room there are magazines and other publications from Europe, Asia and South America. And in the bookshelves are texts on foreign languages. Keplinger has taken up the study of other languages partly

as a hobby, and partly because he likes to converse or correspond with his business associates overseas in their native tongues. . . .

This room, with its floor of pink terrazzo with a sunburst pattern in the center, was Keplinger's favorite, and it was the center room for partying. Sliding glass doors opened on to the patio and backyard.

When the Keplingers entertained foreign guests, it seemed that the Tulsa newspapers regularly carried stories and photographs of the event: "Mr. and Mrs. C. H. Keplinger, 2213 E. 27th St., entertained with a party in their home Saturday night honoring their guests, Mr. and Mrs. Francisco Puyana and Enrique Aparicio of the Colombian Government Oil Co. . . ." And: "*Alliance Française* of Tulsa was honored recently by a visit from Georges A. MacClenahan, Consul General de France, and Jacques Dessourdres, Cultural Attaché de France. Mr. and Mrs. C. H. Keplinger gave a cocktail party for them and their wives in their home at 2213 E. 27th St. . . ."

As much as they enjoyed their home, the Keplingers spent many of their social hours at the Southern Hills Country Club, which a newspaper columnist described as "a second home for Tulsa's 400." Keplinger became a member several years after he launched his own company.

There were parties galore at the club which the Keplingers attended and occasionally hosted. These affairs were grist for the newspaper columnists' mills. One columnist noted:

> It took two red carpets Friday night to accommodate Tulsa's "400," which was much nearer 800 to 1,000 at a quick 'casing' of two major spots. Louise and Henry Keplinger and Catharine and Harry Horton stood in a receiving line at the door of Southern Hills Country Club from 6 to 9 p.m. greeting their friends who were as brightly dressed as the May pole which centered the ballroom. Louise and Catharine each selected blue frocks—Catharine's was a beautiful shade of periwinkle in chiffon and Louise's was a "little girl" dress of light blue with rows and rows of ruffles on the skirt. . . .

Perhaps the Southern Hills golf course was the chief attraction for Keplinger. As pointed out earlier, he learned to love the game in his caddying days at El Dorado and played at every opportunity as he grew older. He carried a 16 handicap onto the links, but he could boast of two holes-in-one over the course. His first was on August 6, 1976, with an 8-iron shot on the par three, 150-yard No. 6 hole. His second was recorded about three years later, on the fourteenth hole, generally considered to be one of the toughest holes in America. His happy grin on that occasion masked the pain which at the time was racking his slender frame. (Louise had shot a hole-in-one on September 19, 1972, the ace coming on the 101-yard eleventh hole.)

Earlier, in 1971, Keplinger won the W. K. Warren Seniors tournament, a Southern Hills event for members over 50 years of age.

One of Keplinger's favorite golfing companions was Bill Graham, a Wichita oilman who had grown up in El Dorado with Keplinger. Keplinger tried to participate every year in the Southern Hills "Swingeroo" tournament, and he enjoyed Graham as a partner. Graham had a fine sense of humor; one year he had a trophy made up—before the Swingeroo started—that proclaimed Graham and Keplinger as winners of the event. "It's the only way we'll ever get a Swingeroo trophy, Henry," Graham said.

Over the years, Henry and Louise entertained many couples at the Swingeroo. In the early days the event provided lavish brunches, lunches, and parties where both members and guests came dressed in costume golf attire. One year Keplinger and his golfing partner, Eddy LaBorde, showed up in authentic kilts from Canada.

Bill Graham was responsible for another "trophy" that graced the Keplingers' clubroom. The Keplingers, Graham, and his wife Marge had spent a golfing vacation in Acapulco. The Grahams planned to stay on several days after the Keplingers left for Tulsa. They were going to drive the Keplingers to the airport. Outside their villa, just as they were about to get in the car, Graham said, "Hit one last ball, Henry."

Keplinger dropped a ball to the ground and drove it toward the golf course. But the ball never arrived; it sailed up into a tall palm tree and didn't come down.

Several weeks later the Grahams came down from Wichita to visit the Keplingers. Graham handed Keplinger a box. Keplinger opened it. Inside was a hand-made scene showing Keplinger swinging a golf club in a tropical setting which featured a tall palm. And nestling in the top of the palm was a real golf ball—the one Keplinger had hit in Acapulco. Graham had hired a villa busboy to climb the tree and retrieve the ball!

One of Keplinger's favorite courses was at the Everglades Club in Palm Beach, Florida, an exclusive club he joined when he and Louise began vacationing in the Florida resort city. But he played everywhere he went. Most of his Southern Hills friends were not aware of this because Keplinger kept his silence about it, but if there was a golf course near any of his business activities, he was certain to play a round or two.

After his company opened its Houston office in 1966 with Kep in charge, Keplinger played golf in that city with his old University of Tulsa classmate Sequoyah (Squaw) Brown. Brown was vice president and director of Exxon Pipeline Company until his retirement and later ran the Houston office of Morgan Stanley and Company, Incorporated. He had been a year ahead of Keplinger at the school and had been elected president of the Student Council, a post Keplinger had vied for the following year but lost. In 1973, both men were honored with Distinguished Alumni Awards by the TU Alumni Association.

Brown also loved golf. "When Henry came to town," Brown said, "it was time for me to get out my golfing shoes."

Keplinger once noted in his work books that he had played golf near every major oilfield in the world where there was a course. This may have been an exaggeration, but in his work books his notes seem to bear him out. When he went to Tierra del Fuego to study the geology for Tenneco Oil Company, he bemoaned the lack of a golf course "at the bottom of the world." The southernmost tip of South America, he noted, was without any amenities, and reminded him of "fifty miles in any direction from El Paso, Texas. . . ."

☐

Keplinger played another role at Southern Hills. It was not unusual for a younger member to approach him on the golf course for a brief conversation. Not even to Louise did Keplinger ever reveal the nature of these visits. But years later one young member after another would tell her that Keplinger had eased his mind about some personal problem, and she would recall having seen him approach Keplinger on the golf course.

It was not only on the golf course that these sober visits were held. Younger people, their brows furrowed with worry, dropped by his office or his home and even stopped him on the street. For each he had a portion of his time.

These visits are not to be confused with the almost daily brief meetings where his peers sought business help or advice or simply wanted to bask in the sunlight of his smile. (One associate said: "I had to quit going to lunch with Henry. So many people would stop him on the street to talk to him I almost starved to death. It was like walking along the sidewalk with the President.")

What was it about this man that prompted people to seek him out with their most personal problems? And how did he always seem to have on hand the words to soothe them? He was too much the pragmatist to suggest prayer as the only hope; the work books are evidence of that. No, he had some gift of spirit that allowed him in the briefest instant to *understand* the peculiar problems that bombard the human heart and soul . . . to *feel* them in that briefest instant as his own. And understanding and feeling them as his own, his keen mind was a jeweler's eyepiece seeking the flaws in reason, hunting sensible answers to mankind's conduct, finding the solution, perhaps, within the varicolored framework of his own existence.

In his work books are quotations, not always credited, from Montaigne to Homer, Pascal to Shakespeare. But the one which appears to be most fitting to these personal visits comes from Judge Learned Hand: "Courage, my friends! Take heart of grace. The devil is not yet dead. . . ."

CHAPTER ELEVEN

E bullient Jim Schlesinger—charmer—hypnotizes his associates—office is spacious, well-used large room with comfortable chairs and lots of reports—Dr. Jim's flat-topped desk was stacked with papers—turns on his charm—voice mellow and full throated. . . ."

The date was November 11, 1977, and in his work book Keplinger was describing Dr. James R. Schlesinger, Secretary of Energy in the Jimmy Carter administration. He had met with Schlesinger on that day in Washington, D.C., in an effort to bring his knowledge and reputation to bear on the nation's future energy policy. It was something he had been trying to do since the 1973 embargo on oil shipments to the U. S. by the oil-producing Arab states.

For years before that he had cried out—in vain—that importation of foreign oil and the resultant lack of domestic exploring, if continued apace, could place the United States in a precarious position—economically, politically, and militarily.

And even before that he had decried the federal regulation of natural gas and had argued strenuously that the price the product brought on the market was "ridiculous to the point of being criminal."

He was a staunch conservative. He voted for a Democrat only when he was convinced that the Democrat was more conservative, or a stronger supporter of the oil and gas industry, than his Republican opponent. But the first person at the seat of power to ever really lend him an ear was generally regarded as a foe of the industry and an "egghead" at that—the same

136

James Schlesinger who, for a time, had the ear of President Carter.

President Dwight Eisenhower had vetoed a bill which would have deregulated natural gas. At the behest of domestic producers he had imposed first voluntary then mandatory limitations on oil imports, but had failed to enforce them, and the flow of oil from abroad continued unabated. Richard Nixon had promised much but delivered little to the domestic producers, and Gerald Ford had done even less. Both Nixon and Ford were in office after the Arab embargo, but Nixon was too engrossed in the Watergate coverup to deal effectively with the energy crisis, and Ford signed a compromise energy bill which contained provisions for a continuation of oil and gas controls and a rollback in domestic oil prices.

Still, Keplinger continued to fight for decontrol and higher prices for domestic oil and gas, and it led him to this confrontation with the "ebullient" Schlesinger. For more than an hour he used all the facts at his command, all of the persuasiveness he had developed after decades of negotiations, to support his basic contention that to reduce the balance of payments for foreign oil there must be adequate incentives for domestic oil and gas development.

Schlesinger wanted more data. Keplinger returned to Tulsa and prepared a report which he sent to the Secretary. The report showed that the replacement price of crude was $4 a barrel more than the price allowed producers by the government, and that the replacement price of gas was 50 cents per Mcf more than the government-allowed price to producers. Correspondence between the two men continued, with Keplinger sending Schlesinger reams of data on every aspect of the energy situation and even praising the Secretary on occasion for some public utterance or what appeared to be a change of opinion.

Keplinger told the *Oil and Gas Journal*: "Dr. Schlesinger is very cognizant that in order to develop the maximum oil and gas reserves the price must equal to the replacement cost."

But Schlesinger was not long in the Carter administration. Keplinger said Schlesinger's willingness to act on energy matters was restrained by President Carter.

"The crux of the whole problem is that Carter is misinformed," Keplinger told the publication:

Due to . . . those who say we're running out of energy regardless of price, Carter doesn't realize that energy supply is price sensitive. The administration predicts no future increases in natural-gas production regardless of higher wellhead prices.

The Carter energy program is unworkable and contrary to the best interests of this country. The consumers of America deserve an energy program designed to increase our oil and gas reserves, not a program which will stifle the initiative to explore and produce. . . .

Keplinger had not always been this harsh in his criticism of President Carter. Carter had presented the general outline of his energy program on April 20, 1977. The oil industry did not respond to it with great enthusiasm. But Keplinger—according to his notes—appeared to hold a degree of faith that Carter would do what others before him had failed to do. When the Tulsa *World* asked for his comments on Carter's program, he responded with this open letter to the President:

I support the goals of your energy program, and the nation will be forever grateful of the nonpartisan program proclaimed by you on the evening of April 20, 1977. However, there is one "bad apple" in the barrel of good proposals which you have given us: failure to plan for reduction of foreign oil imports. You recognized the problem when you stated that the cost of foreign oil last year was $35 billion—a horrendous outgo of money for the United States—but you did not come up with an acceptable solution.

I humbly submit that you must make one correction in the otherwise well-conceived program to make America strong economically and politically. You spoke and America listened and will act. Programs to conserve energy will strengthen the nation's energy base, and all citizens welcome reluctantly the opportunity to buckle down and help. I cannot say enough good things about your energy proposals, but your plans will be doomed to failure without one change in the plan—to effectively reduce foreign oil imports and to remove us from the domination of OPEC (Organization of Petroleum Exporting Countries), a foreign group of oil countries supplying us with increasing volumes of foreign oil.

Your proposal to reduce foreign oil imports to 6,000,000 barrels per day has to come from a buildup of our domestic oil and gas production. To accomplish your reduction in oil imports you must amend your program to "regulate the price" of new oil and gas discovered in the future, after April 30, 1977, to an amount equal to the equivalent cost of foreign oil.

The Federal Energy Administration must be given the legislative power by Congress to set the price of new oil and gas equal to the exact costs of foreign oil on a monthly basis—if foreign oil costs go up, FEA must have the power to increase the price of new oil and gas produced in the United States to make us competitive with the OPEC-regulated administered world oil price.

Regulation of the new oil and gas price by the FEA would reduce the uncertainties for the oil and gas producers and allow them to make plans to discover the large potential new oil and gas reserves of the United States to replace the purchased foreign oil.

Regulation of new oil and gas prices would make the Carter Energy Plan workable. Otherwise, without incentives to replace foreign oil, we cannot by conservation alone do the job required to strengthen our energy supplies.

Under the proposed regulation feature, the new oil and gas discovered immediately starts to replace foreign oil. We must have this regulation so that the OPEC nations will not continue to control our destiny. We have governmental machinery to regulate the new oil and gas prices in the FEA today.

Keplinger's notes reveal that he considered Carter a "good man with good instincts." One jagged scrawl across a sheet of paper said, "Carter is trying." But as the months passed he became impatient, and his public utterances betrayed it.

By now he had found a forum, a supporting group—the Energy Consumers and Producers Association (ECPA) of Seminole, Oklahoma, "an interested group of U. S. citizens who believes in the ability of the petroleum industry to find domestic oil and gas reserves to replace the purchase of foreign oil. . . ."

The group first engaged him as a consultant. Then he became a member and later the organization's chairman.

Keplinger went to Washington as the group's spokesman on May 24, 1977, to testify before the House Ways and Means Committee. He swamped the legislators with graphs, charts, and other material while he pounded away on the theme he had employed in the open letter to Carter. The President's suggestions to combat the energy crisis were "laudable." "Conservation programs," Keplinger said, "are rewarding, but they do not increase our dwindling oil and gas reserves."

What his group proposed, he said, "is that the oil price from new oil wells would be 'regulated' and not allowed to exceed the United States' delivered price of foreign imported oil, which cost us over $35 billion last year and is estimated to exceed $40 billion this year.

"This 'regulated' price for new oil wells would be the same as is now being paid for imported oil, so there would be no 'oil rip-off' for the consumer. In addition, the new wells in the United States would be drilled by American labor and equipped with materials produced by American labor in all sectors of the nation.

"The consumer would pay the same price for new domestic oil as for foreign imported oil, but there would be no outgo of United States dollars to purchase foreign oil. The development of our domestic oil potential would generate work and business which would create additional federal and state taxes. This would strengthen the dollar and the economy. The only loser would be the OPEC nations who would sell us less and less imported oil as our domestic drilling increased. . . ."

What Keplinger *really* wanted, what the oil and gas industry—majors and independents alike—*really* wanted, always had wanted, was to go about their business without any government interference at all. They wanted oil decontrolled and gas deregulated, arguing that the American consumer would profit in the long run. The independents had presented this argument for decades. But the energy crisis had prompted them to modify their stand. For the American public laid the blame for the crisis at the door of the oil industry. The 1973 embargo had come as a shock, but no more so than the sudden increases in prices for petroleum products and the profits the major companies made from the increases. No explanation the companies offered seemed to appease anyone. Even in 1977, when Keplinger appeared before the House committee, oil and gas were still volatile political mixtures, and the Democrats controlled both the legislative and executive branches of government. The most Keplinger and his group could hope for, it appeared, was some kind of control phase-out over a period of years.

But they fought on during that summer, trying to influence Congressmen, and that fall Keplinger moved his attack to the executive branch with his visit to and correspondence with Secretary Schlesinger. All through 1978 he breathed optimism that energy legislation was forthcoming which would put the U. S. on the path to energy self-sufficiency. While in Palm Beach, he told the *Daily News* on March 30: "From what I've seen, I'm very encouraged. They [legislators] are making an attempt to please not only the oil men but the consumers as well. . . ."

From the depths of his experience he dredged up the data to substantiate his urgent appeals for rapid exploitation of coal, oil shale, and solar power for the long-term while prodding the administration to accept as fact the immediate need to "unchain" the domestic oil and gas industry.

When OPEC announced an unexpectedly steep increase in the price of crude in late December 1978, Keplinger wrote another open letter to President Carter. He repeated all of his arguments, which by now were well-worn, and said: "A new day can break if we will follow an energy policy of immediately allowing world oil prices for every barrel produced from new wells completed after January 1, 1979."

By now Keplinger was laboring under a physical handicap that probably would have daunted less dedicated men. But he hid his pain and discomfort beneath a mountain of work.

The President's long-awaited "energy package" was slated to be revealed on April 5, 1979. But well in advance of the date the contents of the package were being publicly discussed. The Tulsa *Tribune* reported: "Battling down to the eleventh hour, the Oklahoma-based Energy Consumers and Producers Association buttonholed influential national leaders in Washington last week to push for decontrol of crude oil prices—ahead of President Carter's address on the subject to the nation this week. Representing the Seminole-headquartered organization was Tulsan C. H. Keplinger. . . ."

The story ran on April 2. On April 5, and later on April 26, President Carter announced the phased decontrol of oil prices beginning June 1, 1979, and ending with the expiration of price control authority on October 1, 1981. And the President proposed a "windfall profits tax" to "prevent unearned, excessive profits which the oil companies would receive as a result of decontrol and possible future OPEC price increases. . . ."

The tax money, the President said, would go to provide assistance to low-income households who could least afford energy price increases; to increase funding for mass transit; and to undertake a major program of new energy initiatives and investments which would permit the country to develop critically needed alternatives to imported oil.

So the battle began anew. And while it was at its height, revolution broke out in Iran, the Shah fled the country, and Iranian oil ceased to flow to the U. S. and other western nations. Keplinger immediately suggested that producers be allowed to drill in present oilfields in small space drilling locations, a move which would in some areas double the number of producing wells. Water injection also would force out more oil, he said. But he held little hope that he would be heeded.

The focus of attention now was the windfall profits tax, which was being debated in Congress. Keplinger and the ECPA called it a "punitive federal excise tax on oil production." Keplinger said: "The energy industry has no quarrel with an excess profits tax as in World War II. However, we adamantly disagree with the windfall profits tax because it is an excise tax and has nothing to do with profit. It merely takes capital from the producer and gives it to the government. It has no relation whatever to investment plans or opportunities. It has no relation to production potential. Lack of capital will prevent the industry from drilling more wells in the 1980s. Drilling in 1978 and 1979 were nearly the same. Why doesn't government encourage industry to drill at the increased rates of the 1950s?"

The fight wore on until March 1980. While the tax bill was in its final stage of passage, Keplinger and more than 600 believers, wearing hard hats and laden with oilfield equipment, including a work-over rig, marched on Washington. They set up the equipment across the street from the White House, and Keplinger "starred" at a press conference.

No one but his most intimate friends were aware of the toll the fighting was taking on the reservoirs of his strength. He told the news media: "America needs more domestic oil production, not taxes, at this time. The geological exploration teams throughout the United States are encouraged that we have a large potential of new oil producing capacity, but with inflation and the high risks involved to find new oil, we respectfully ask Congress not to put the 'windfall profits' tax on the new oil category. In the final markup of the bill, the Congressional Conferees have taxed the future new oil which we hope will be found and, in our opinion, as a result will decrease new domestic oil found in the next ten years by 25 percent. This can only lead to greater dependence on importation of foreign oil. . . ."

And he invited the reporters—and all of Washington, for that matter—"to see this actual display [of equipment] of what is used in the oil fields. . . ."

He walked the halls of Congress, trying to talk to any legislator he believed would have influence on the bill's passage. In a two-day period he visited every member of the Senate Finance Committee. He had believed that two Democratic senators, John Glenn of Ohio and Henry Jackson of Washington, might be persuaded to his view. He was wrong. The tax became effective.

Keplinger supported George Bush against Ronald Reagan in the race for the Republican presidential nomination. After Bush lost, Keplinger supported Reagan against Carter in the general election. Reagan won by a landslide, and shortly after he took office his administration lifted the controls on oil.

But he took no immediate steps toward repealing the windfall profits tax on oil, and he postponed action on the decontrol of natural gas prices.

Reagan's failure to act on the two issues would not have surprised the pragmatic Keplinger. About the windfall profits tax, he told an interviewer shortly before his death: "It will be hard for any administration, Republican or Democratic, to give up all that money once they get their hands on it." About the de-control of natural gas, he said: "I don't care what kind of proposal Reagan makes to speed up gas decontrol, it won't be approved by Congress without a surcharge on gas revenues. I hope I'm wrong, but I think I'm right. . . ."

CHAPTER TWELVE

I n 1976 and again in 1977, Keplinger and Louise vacationed in Spain. They spent those holidays in a charming villa near Marbella on Costa del Sol. From there they visited the Alhambra and other points of interest. They had been offered the villa by a Tulsa entrepreneur who held Keplinger in high regard.

The villa was on a hill overlooking the sea on one side and a golf course on the other. It was a land of gorgeous sunrises and sunsets, warm days and star-filled nights. The couple was alone except for Pepe and Antonia, a cook and maid who lived and worked year-round in the villa. Keplinger, of course, began teaching Pepe English while putting a fine polish on his own Spanish.

Before they left Tulsa for Spain on the second trip, the entrepreneur told Keplinger that he had ordered some repair work on an eroding retaining wall. "Look into it, if you please," he suggested. Keplinger said he would.

The wall was eight feet high and badly worn in places. "Henry couldn't restrain himself," Louise said later. "The first thing I knew he was supervising the job, telling everybody what to do and how to do it. He enjoyed himself immensely."

(Louise probably was surprised, too. Keplinger was not "handy" with tools. Once in Centralia he and a friend built a flat-bottomed boat in a single day. The next day, while Louise and Keppy watched expectantly on the shore, the men proudly set their handiwork afloat in a nearby lake—and it promptly sunk.)

Keplinger was enchanted by Spain and the Spanish people.

He told Louise, "I want us to come back here every year for the rest of our lives." Louise agreed.

So they were planning to return to Spain in 1978 where they would celebrate his birthday and their wedding anniversary. He was 68 that year, feeling good and playing in the Swingeroo at Southern Hills, working very hard in Washington for the things in which he believed, working elsewhere at his business, which he loved as much as in the day when he first started on his own. His children, Kep and Karen, were grown, with families of their own. He had told Louise when he proposed to her that he had a lot to accomplish in his life and that he wanted her by his side. He had accomplished much, this boy from the Kansas prairie, and always she had been where he wanted her to be. She always would be.

He went to his doctor because he felt a bit dizzy. The doctor plucked a bug from his ear. "That's why you've been dizzy," the doctor said. "But look, you're two months overdue on your annual physical. Let's have a look at you."

"When I get back from Spain," Keplinger said.

But the doctor insisted. "Right now."

So Keplinger submitted to the examination. Everything appeared in order.

But the doctor called him a few days later. "Henry, come over and take one test again. I want you to do it."

"No," said Keplinger. "I've got to play this match in the Swingeroo. If I don't, I'll have to forfeit."

"I'm concerned, Henry. It could be a mistake, but I'm concerned."

So Keplinger went.

His work book pages, usually so full of his visits and other activities, bore only two words on July 14: "Liver Cancer."

He already had written the word "Spain" at the top of each page of his work book through August 7, with "Spain return" written on August 8. But he ran a line through the words, and those pages now were devoted to his initial confrontation with his dread enemy.

His family and loving friends had arranged for him to enter the John Stehlin Cancer Research Center at St. Joseph's Hospital in Houston. In an essay he wrote for his family, friends, staff, and clients, he described his sustaining faith. The essay was entitled, "God's Celestial and Mortal Angels." It said:

> I have recently had a health problem which reaffirms God's power to provide Christians with both offensive and defensive weapons through his heavenly host of angels who carry out God's wishes as messengers to all Christians in the world, and God's mortal angels, you and I, who give pragmatic help to God's followers.
>
> Recently I have sensed the presence in my life of both celestial and mortal angels. Throughout the history of the bible, God has used angels as His messengers to carry out His judgments. God's angels constitute an infinite body of helpers from Archangel Michael, Gabriel, and the seraphim and cherubim.
>
> I had planned a vacation for June, 1978, but was hindered from going due to roadblocks which caused me to postpone leisuring [his work in Washington]. I believe God, through his angels, stopped me because I had a pending health problem. I had *no* pain and felt that I was in perfect health.
>
> Finally, I had my annual physical examination in which my doctor, through blood tests, discovered something malfunctioning in my body. He immediately did a series of tests which indicated I had a tumor on my liver which was quite large and diseased and should be removed. I had no pain or discomfort, but this would not last for long and the problem had to be corrected.
>
> At this point, God's mortal angels went to work to determine the most likely clinic which could solve my problem. Within 24 hours, a host of God's helpers, as you and me, had researched the field and *I* was in Houston, Texas, at the John Stehlin Cancer Research Center. God had chosen the physician to solve my problem.
>
> Doctor Stehlin and his magnificent staff went to work to prepare me for the operation and get the details of my problem.
>
> On the eve of my operation, Sunday, July 23, 1978, I sat cross-legged on the bed visiting with the doctors, Dr. John Stehlin and Dr. Peter Delpolyi, without a pain in my body. Through a liver scan they showed me an area 4 by 4 inches covering a portion of my liver. I was advised it was a most

serious operation and they wished me to consider the seriousness of the operation.

I was with the doctors and all of my immediate family. I turned to God for His blessing and help, and in a fraction of a second He said: "Do it." I had peace of mind. I told Dr. Stehlin that I was ready.

The following day the skilled medical team, with God's help, operated and everything worked—just like a miracle being performed only for me.

Christians have God's messengers, both the celestial and mortal angels, to solve their health problems. His mortal angels are skilled physicians and all others who aid in curing the ills of God's followers.

For a patient, get-well cards, letters, calls, pretty flowers and plants, and books aid greatly in promoting patient morale. Those who act in this capacity, particularly blood donors, are truly a part of God's mortal angel corps. Above all, the love and care of your own family provide healings which are most essential.

Even after nine-and-a-half hours of surgery, his morale should have been high: flowers were banked against the walls outside his large room, and the room itself was on the verge of overflowing with colorful plants and blossoms. For decades Keplinger had been cheering up others; now the get-well cards and messages flowed into the fragrant room from cities around the world.

And all of the nurses just *knew* he was Senator Barry Goldwater of Arizona. They whispered about it in the hallways, and Keplinger, with a straight face, did nothing to deny it. For years women had often mistaken him for Senator Goldwater. Some seemed disappointed when told that he was just plain Henry Keplinger of Tulsa, Oklahoma, so to save their feelings he had adopted a plan of neither affirming nor denying their contention. The nurses became even more certain that their charge was Goldwater when they found Keplinger and Louise dancing in the room, Keplinger with several tubes running to his body. Only a romantic like the handsome Senator would caper like that!

☐

Once out of the hospital, Keplinger sailed right back into his busy work schedule and into the thick of the energy fight in Washington. In-between he sandwiched a trip to China as a guest of the Chinese government. He told an interviewer: "After that trip I felt like I had been everywhere a man could go. I felt like I had seen it all."

He flew to China via San Francisco, Okinawa, and Hong Kong. (He attended the Rotary Club meeting in Hong Kong.) From Hong Kong he rode a train to Canton, and he noted in his work book, "Great train ride!" From Canton he went on to Peking where the congregated Chinese geoscientists welcomed him as first among peers. "I was pleased that they had heard of my work and had read my papers," he noted in his work book.

He visited oilfields. He studied material the Chinese allowed him to study, and he was frank in his comments on their problems and the country's potential. And he visited the Great Wall and the Ming Tombs. "Fascinating people, fascinating artifacts, unlimited future if. . . ." he noted.

Perhaps it would be inferring too much to say that Keplinger loved all mankind. He was an elitist, in many respects. But he told an interviewer, "I've come close to seeing every kind of human being that lives on this earth, and the only basic difference I've noticed is that we all don't look alike."

Shortly after his return from China, Keplinger went to Houston to have his doctor examine a pimple on his body which was annoying him. A biopsy showed it was malignant. A liver scan revealed that the liver was affected.

After exploratory surgery, the physicians decided to treat Keplinger by way of a two-pound pump which would be strapped to his body. The pump would release medicaments which would drip through a tube to his liver.

While this mobile device was being readied, Keplinger was "hooked" to a stationary machine which pumped the medicaments to his liver. He would be "on" the machine for two hours, then "off" for two hours.

During this hospital stay he became accustomed to listening to Dave Fowler on radio station KPRC. Fowler was the most popular talk-show host in the city, an intelligent, knowledgeable man who knew when to be brusque and when to be compassionate with the listeners who phoned in to talk with him. And he liked to have an occasional guest on the show who could answer questions on a particular subject.

Fowler was going through a trying time. Day after day his callers seemed obsessed with the energy situation, trying to understand it and trying to understand what was going on in Washington. He had some prominent oilmen on the show, but the answers they gave the callers apparently were not satisfying.

In his hospital room, "on" or "off" the machine, Keplinger listened to Fowler's show with growing exasperation. Louise had a bed in the room as usual, for she seldom left his side. "I'm going to get on that show," Keplinger told her—and he would not listen to argument from anyone.

He called his Houston office, and someone in the public relations department got in touch with Fowler and sent him Keplinger's imposing resume. Fowler said later, "The resume was impressive, so I said I would like to have him on the show."

Keplinger had two hours in which to get to the station, be on the program for an hour, then get back to his hospital bed and the machine. Louise went with him to the station.

Said Fowler: "Frankly, I had never heard of the man, but I'll never forget him. He answered every question put to him, by me or by those calling in, clearly and concisely. He had that ability to explain complex matters in simple terms. When he was through, I felt like I thoroughly understood the energy situation in all its aspects, and I know the listeners were pleased with him." Fowler smiled. "You know, I wasn't aware that he was sick until he started to leave. He waved a hand and said, 'Well, Mr. Fowler, I have to get back to the hospital.' I realized then how important it had been to him to explain and answer questions on the show. He figured he was doing a public service in helping to educate his fellow citizens."

□

Keplinger left the hospital wearing the two-pound pump strapped to his body. Whatever its merits, the pump was a major annoyance. Keplinger couldn't sleep. He couldn't take a shower. He became nervous, and his stomach bothered him.

Perhaps Keplinger was not fully aware of the toll his labors were taking on his failing strength. He wanted to continue his business and political affairs as usual. Louise was torn between catering to his wishes and trying to preserve him. Every time a pain swept through Keplinger's body, it swept through hers.

They were in Butler, Pennsylvania, Louise's hometown, when Keplinger became desperately sick at his stomach. He was attending a board meeting of Spang & Company, of which he had been a director for many years. With help, Louise got him to Pittsburgh and on a plane to Houston. He had just been received at St. Joseph's Hospital when an ulcer in his duodenum erupted and he began vomiting blood.

Doctors worked for a week to get the bleeding under control—and again the ulcer broke. This time he underwent emergency surgery. The six-hour operation was successful, but he was slow to recover. Nevertheless, after coming so close to losing his life, he went back to work, alternating his time between Tulsa and Palm Beach, where he and Louise had an apartment. Among other things, he was making some studies for Dome Petroleum Corporation, and he would dictate his reports long distance from Palm Beach to Anna Brown in Tulsa; she would transcribe the shorthand notes and send the reports to Dome.

Keplinger would not refrain from playing, either. On a fishing trip aboard Kep's boat, *Sukeba II*, he caught an eight-foot sailfish. Louise, Karen, Kep, and Louise's mother were on the trip, and if anyone was more thrilled by the catch than Keplinger it was Mrs. Spang. Adding to his pleasure was this: a sailfish derby was being held out of Palm Beach at the same time, and the winner's fish was considerably smaller than Keplinger's trophy. His illness could not diminish his triumph.

And he found the time to join four other energy experts in contributing to a Wall Street Journal advertisement for the LTV Corporation. The others were Michel T. Halbouty, a Houston-based geologist, petroleum engineer, and independent producer; George Gaspar, vice president and board member with the investment firm of Robert W. Baird; John F. Bookout, president of Shell Oil Company; and Gene T. Kinney, editor of the *Oil and Gas Journal*. The advertisement was headed: "The Energy Finders: They've Searched Around the World, Under

Land, Sand and Sea. Where Next? LTV's Continental Emsco Is in Energy; We Asked 5 Industry Experts to Look into the Future."

Keplinger's section of the advertisement was entitled, "A Look at U.S. Supply and Demand in 1990." It said:

> One of the greatest advances of the 1980s will be to include the close cooperation of the energy industry with the Administration and Congress.
>
> Every effort must be made to open up and expand the exploration areas in the United States which have been withheld by the Federal government for expanding the wilderness areas. The nation needs to have multiple land use whereby we can explore for the minerals and the oil and gas and, at the same time, have a satisfactory environment.
>
> We don't know precisely what the domestic production of oil and gas will be in 1990, but our best estimates are shown in this chart. [The chart showed that the total oil supply equivalent without foreign supplies would be 39,900,000 barrels per day, and that total U. S. demand would be 47,200,000 barrels daily.]
>
> The key number is the 39,900,000, because when we look at the demand picture for 1990, we arrive at an estimate of 47,200,000 barrels per day—oil equivalent.
>
> The difference between the demand and supply is that we will require 7,300,000 more barrels of equivalent oil per day from foreign imports. This is the bad news for the energy picture in the 1980s.
>
> It is my hope that future legislation will be undertaken which will give more incentives to produce more domestic oil and gas, coal, nuclear power, synthetic fuels, and hydroelectric and geothermal power so that this estimate of rather large imports of foreign oil can be revised downward significantly.
>
> In conclusion, we need a balanced energy program to produce our abundant energy supplies under reasonable environmental conditions, and above all, each American must conserve the use of energy.

The essay was his last public cry for the things in which he believed, the last attempt to shape the energy future of his beloved country. For another liver scan had revealed a recurrence of the malignancy, his general health was poor, and his duodenum was bothering him again.

His spirits were lifted, however, when his comrades-in-arms at Energy Consumers and Producers Association joined with others in the industry from coast to coast and border to border to honor him as ECPA's "Oil Man of the Year." The affair was held in Tulsa's Camelot Inn on June 12, 1980. Keplinger in the past had received dozens of awards, plaques, and scrolls from various engineering societies and governmental bodies, but it is likely that none thrilled him as much as the ECPA honor.

E. L. (Bud) Stewart, ECPA executive director, told the Tulsa *World*: "This doesn't mean the award will be an annual thing, because you don't run into Henry Keplingers every year."

Emaciated by his illnesses, his hair almost gone due to chemotherapy, Keplinger somehow found the strength to stand straight and proud while Michel T. Halbouty, the chief speaker, and others lauded him to the Camelot ceiling. "He was so sick," Louise said. But he was buoyed by the affection and regard that floated across the great room like incense of the Magi.

□

Later that year, in October, a story appeared in the Tulsa newspapers about Dr. Stanley Order, Director of Radiation Oncology at Johns Hopkins University in Baltimore. Dr. Order, the story said, was employing a revolutionary procedure in the treatment of tumors. Shortly thereafter, friends and relatives called members of the Keplinger family with word of Dr. Order. Keplinger was worsening. Kep called Johns Hopkins and learned that Dr. Order was in Galveston to speak to a group on his method of treatment. Kep called him in Galveston. He had learned that Dr. Order liked to fish, so he told Dr. Order that he had a boat docked at Galveston. "You can examine Dad there," Kep said. "Fine," said Dr. Order.

Keplinger and Louise flew down to Galveston with all of his medical records. On the boat, Dr. Order studied the medical records, then examined Keplinger. He decided that he was a candidate for the new treatment. But first, he said, Keplinger should receive some primary radiation locally.

Keplinger was heartened by Dr. Order's decision, and the family left Galveston, leaving Dr. Order to fish on the boat or cruise Galveston Bay as he wished.

Keplinger spent two weeks in M. D. Anderson Hospital in Houston undergoing initial radiation, and later went on to Baltimore for the Order treatment at Johns Hopkins. The results of the treatment exceeded all expectations, but Keplinger contracted sepsis. So, while the radiation treatment was deemed most helpful, his body was ravaged by the sepsis and the massive doses of penicillin he took to curb it.

That was in December 1980. He left the hospital frail but convinced that he would whip the cancer. He went back to Baltimore in January, and again in February, for checkups. The doctors were pleased with his progress and, in both instances, decided against further radiation treatment for the time being. They told him to come back in March for another examination.

Keplinger's jubilation over the results of the cancer treatment was tempered, however, by a constant, debilitating pain, and he was extremely ill when he and Louise returned to Baltimore in March.

In their hotel, before Keplinger was to go to Johns Hopkins for his examination, he started hemorrhaging. The indomitable Louise got him to the hospital. Once again Keplinger underwent emergency surgery to halt the heavy bleeding.

He survived the surgery, but the well of his strength had been depleted. Still he clung to life until his courage and iron will no longer could sustain him. Louise, Karen, and Kep had been with him at every crucial stage of his long illness. Indeed, it seemed as though Louise had spent as much time in the various hospitals as had Keplinger. So they all were with him in his final days, comforting him as best they could. He died peacefully at 8 p. m. on April 19, Easter Sunday.

☐

There were no assigned pall bearers. He had so many friends, so many who loved him, that Louise decided not to choose among them. Instead, funeral home attendants served in that capacity.

Funeral services were held in the First Presbyterian Church in Tulsa, with Dr. William Wiseman officiating. Keplinger, a deacon, had worshipped in the magnificent church for decades, and Dr. Wiseman had known him as a brother in Christ. So many came to pay their last respects that they overflowed the chapel. Hundreds across the country who could not attend for one reason or another wired or wrote their regrets and expressed their love for their old friend and extended sympathy for the bereaved family.

He was so much a part of Tulsa that it seemed as if the city bowed its head and sighed at the loss. On the street in front of the church an aging oilman with tears in his eyes said to a companion, "We'll never see his like again. . . ."

The far wanderer, the "engineer's engineer" with the heart of a poet, the consultant who placed integrity above competence and knowledge, was laid to rest in a crypt in Memorial Park.

□

"You've set yourself up as a consulting engineer, so be one," Mike Benedum, the great wildcatter had said to Keplinger back there in the late 1940s. "Never sell a deal for a client, and never take one. The most important thing you have to offer is your integrity. Never shade it. If you think a reservoir is holding a million barrels of oil, never let a client talk you into saying it's holding a million and a half. I'm sure you're a fine engineer, but your integrity can make you more money than your talent. . . ."

Keplinger would have lunch on many occasions after that with the old wildcatter, and Benedum would smile and say, "No tarnish, Henry?"

And Keplinger would smile and say, "Not yet, sir."

And there never was.

AFTERWORD

I n the preceding pages we have witnessed my father's life through the eyes and hearts of countless friends and business associates. Many of them are still alive today who saw what is now The Keplinger Companies grow from a tiny two-man office in Tulsa to the world's largest energy engineering firm with 14 offices around the globe and almost 400 employees. The Tulsa office, in the 320 South Boston Building, is second in size only to the Houston office in the Entex Building.

My father suffered the loss of his parents, his beloved grandmother, Nora Suffield, and Ferdinand Spang, his father-in-law and confidant. He cared deeply for them all. But he felt blessed that he lived to see and enjoy six grandchildren. My wife Velma and I produced four of them—Kristine Louise, Karen Anne, Kimber Lee, and Charles Henry II. My sister Karen married an outstanding stockbroker of Tulsa, William Hale Mildren, and they have two children, Marilyn Louise and Matthew Hale.

How did I see my father? When I was young I was conscious that he was away from home a lot. I'm sure that when he returned from a trip he told me stories about what he had seen and done, but like my children are today, I was wrapped up in my own concerns. Although Mother had the chief responsibility in bringing up Karen and me, Dad always took a very active interest in our lives. When he was out of town in this country, he never failed to call at night to see how we were or to talk to us if we were at home.

And as I neared manhood, Dad was there when I needed him. I gave him a gray hair or two, but he let me see the pride he felt when I was graduated from The University of Texas with a degree in petroleum engineering, and later when I took a Masters degree in petroleum engineering and geology at his alma mater, the University of Tulsa.

As his mother had done with him, he taught me the value of hard work. As he had done, I held summer jobs and tried to give them my best. I went to work for him in Tulsa in 1962. Four years later I learned how deeply he was committed to my future.

I had traveled from Tulsa to Houston on several jobs, and I got the feeling that if a man wanted to work hard and accomplish a lot, Houston was the place to do it. I was sitting in the Petroleum Club one night, watching the fog roll in and make halos around the millions of lights, and I could sense the city's electricity. The great energy boom hadn't started, but you could feel it coming.

Back in Tulsa I told Dad that I saw some advantages in operating in a larger city, and that I thought Houston was going to be the center of energy activity in the next 20 years. I told him I thought we should open a Houston office.

I can guess what went through his mind. I was still a bit too carefree in my personal conduct for his liking. I had given no indication of being executive material. I was an untried son, and I was asking him to risk his firm's reputation and money on a shot at the moon.

In his careful way he thought it over, then told my mother, "I'm not going to try to protect Kep from making mistakes. I don't think I should. If he makes them, I hope he can profit from them."

And so he took the chance on me.

☐

As the years passed and the company grew, Dad and I were partners on everything. But in 1970 he decided to turn the management over to me while he concentrated on the work he liked best—the intricate lawsuits with all their drama, involved government contracts, special foreign assignments, and specific jobs where he believed his expertise was particularly needed. Later, of course, he spent much of his time and strength in the fight to shape the country's energy policy.

But while I was expanding the firm through acquisitions and

increased business and building a cadre of top-flight management personnel, I knew I could always turn to him for advice if it was needed. I know he was proud of our thriving organization with its talented management personnel and dedicated employees working in an atmosphere that is pleasant and, we think, inspiring.

Integrity is the rock on which he built.

Integrity is the rock on which The Keplinger Companies will continue to grow because the gifted people supporting me now and looking to the future will have it no other way.

□

On December 14, 1935, Dad became a registered professional engineer in the State of Oklahoma. He was only 25 at the time, the youngest engineer so accredited. Names of many of the organizations and societies with which he was affiliated are scattered through this book, but I think they are worthy of listing in addition to those not mentioned.

He was a registered professional engineer in Oklahoma, Wyoming, and Alberta, Canada; member of the American Society of Mechanical Engineers and the Society of Petroleum Engineers of AIME, charter member and past-president of the Society of Petroleum Evaluation Engineers, and member of American Gas Association, American Petroleum Institute, American Arbitration Association, Independent Petroleum Association of America, Mid-Continent Oil and Gas Association, Tulsa Geological Society, American Association of Petroleum Geologists, Mexican Geological Society, Mexican Petroleum Engineering Society, L'Association Française des Techniciens du Pétrole, NOMADS, and the Oklahoma Independent Producers Association.

Dad was named Distinguished Alumnus of the University of Tulsa in 1973. He was presented the Distinguished Service Award by the Oklahoma Petroleum Council in 1979, was advanced to the Legion of Honor by the Society of Petroleum Engineers of AIME in 1980, and was made an honorary member of the 25-Year Club of the Petroleum Industry. His work for the Energy Consumers and Producers Association resulted in his being named Oil Man of the Year by that group on June 12, 1980. He had been chairman of the association for two years.

He served as a member of both the Energy Resources Committee of the Interstate Oil Company Commission and the Oil Activities Committee and Education Committee for the Tulsa Chamber of Commerce; was a past-chairman International Service in the Tulsa Rotary Club; and held offices in various other civic organizations.

He also was past-president and member of the Tulsa Club and a member of Southern Hills Country Club and the Everglades Club of Palm Beach, Florida.

And Dad was a long-time member of the First Presbyterian Church of Tulsa where he served on three separate occasions as a member of the Board of Deacons—under Dr. Edmund Miller from 1954 to 1957, Dr. Bryant Kirkland from 1964 to 1967, and Dr. William Wiseman, who officiated at his funeral, from 1968 to 1971.

☐

I am grateful to the scores of persons who searched their memories and their files to help make this book possible. I am particularly obligated to Anna Brown, Dad's secretary from 1959 until his death. She knew him as no one outside the family knew him and, I am proud to say, admired and respected him. In turn, she had his admiration and respect. She worked long and hard finding the bits and pieces from which this work was woven. And her personal recollections gave life to sterile facts.

But, as Anna Brown told reporters on several occasions, Charles Henry Keplinger, known by so many around the world, was a very private person. And it was in his workbooks that the inner man was revealed. It was in the workbooks that he scrawled his love of poetry and philosophy, that he jotted down the thoughts he would never utter. The workbooks make it evident that his life was shaped to a great extent by his "visits" with the thinkers of the past.

Across one page he scribbled in 1970 a line from René Descartes, the great French logician and philosopher: *To be possessed of a vigorous mind is not enough; the prime requisite is to rightfully apply it. . . .*

INDEX